'Against a backdrop of increasing food i̲n̲ ̲ ̲ ̲ ̲ ̲ ̲ ̲ ̲ ̲ ̲ ̲ ̲ ̲ ̲ _ɔ countries, Riches' examination of food banking reveals the exte̲ ̲ ̲ ̲ ̲ ̲ ̲ _uch "Big Food" and privatized food charity have well and truly moved into the spaces left by retreating neoliberal governments. Beyond the food drives, celebrity endorsement, smiling volunteers and government legislated tax incentives, this book documents the juggernaut that is global food banking. Despite being a thorn in the side of many "primary duty bearers", rights-based approaches to food offer promise as an effective counterweight to slow the progress of the foodbank juggernaut and reclaim public policy.'

– *Sue Booth, Flinders University, Australia*

'Graham Riches' in-depth analysis of the way food banking has entrenched itself in the neoliberal agenda and public discourse calls for a change in "the conversation about domestic hunger from corporate charity to the right to food". This book makes a significant contribution to this new conversation, arguing that civil society across OECD countries can and should hold the "indifferent States" to account for their failure to ensure dignified access to good food for all when they so clearly have both the means and the duty to do so.'

– *Pete Ritchie, Nourish Scotland, UK*

'Can't we do better than food banks? Graham Riches moves the needle from charity to the human right to adequate food and nutrition. He describes how capital-soaked transnational corporations monopolize public policy, blame poverty on the poor, and endorse themselves as the publically-subsidized solution. Riches' alternative vision, rooted in social solidarity examples, rebuilds the social contract between civil society and its governments through democratically evolved plans, transparent monitoring, and the active participation and leadership of policies' most affected publics.'

– *Anne C. Bellows, Syracuse University, USA, and Board Member,*
FIAN International

'Graham Riches's book is essential reading for anyone wanting to understand why problems of hunger and food insecurity are unabated in countries where food banks become established as the primary response.'

– *Valerie Tarasuk, University of Toronto, Canada*

FOOD BANK NATIONS

In the world's most affluent and food secure societies, why is it now publicly acceptable to feed donated surplus food, dependent on corporate food waste, to millions of hungry people? While recognizing the moral imperative to feed hungry people, this book challenges the effectiveness, sustainability and moral legitimacy of globally entrenched corporate food banking as the primary response to rich world food poverty. It investigates the prevalence and causes of domestic hunger and food waste in OECD member states, the origins and thirty-year rise of US style charitable food banking, and its institutionalization and corporatization. It unmasks the hidden functions of transnational corporate food banking which construct domestic hunger as a matter for charity thereby allowing indifferent and austerity-minded governments to ignore increasing poverty and food insecurity and their moral, legal and political obligations, under international law, to realize the right to food.

The book's unifying theme is understanding the food bank nation as a powerful metaphor for the deep hole at the centre of neoliberalism, illustrating: the de-politicization of hunger; the abandonment of social rights; the stigma of begging and loss of human dignity; broken social safety nets; the dysfunctional food system; the shift from income security to charitable food relief; and public policy neglect. It exposes the hazards of corporate food philanthropy and the moral vacuum within negligent governments and their lack of public accountability. The advocacy of civil society with a right to food bite is urgently needed to gather political will and advance 'joined-up' policies and courses of action to ensure food security for all.

Graham Riches is Emeritus Professor and former Director of the School of Social Work, University of British Columbia, Vancouver, Canada. As author and co-editor he has published widely on rich world domestic hunger, social policy and the right to food including *Food Banks and the Welfare Crisis* (1985); *First World Hunger* (1997) and *First World Hunger Revisited* (2014).

ROUTLEDGE STUDIES IN FOOD, SOCIETY AND THE ENVIRONMENT

For further details please visit the series page on the Routledge website: http://www.routledge.com/books/series/RSFSE/

FOOD BANK NATIONS

Poverty, Corporate Charity and the Right to Food

Graham Riches

LONDON AND NEW YORK

First published 2018
by Routledge
2 Park Square, Milton Park, Abingdon, Oxon OX14 4RN

and by Routledge
711 Third Avenue, New York, NY 10017

Routledge is an imprint of the Taylor & Francis Group, an informa business

British Library Cataloguing in Publication Data
A catalogue record for this book is available from the British Library

Library of Congress Cataloging in Publication Data
Names: Riches, Graham, author.
Title: Food bank nations : poverty, corporate charity and the right to food / Graham Riches.
Description: Abingdon, Oxon ; New York, NY : Routledge, 2018. | Series: Routledge studies in food, society and the environment | Includes bibliographical references and index.
Identifiers: LCCN 2017052875| ISBN 9781138739734 (hardback) | ISBN 9781138739758 (pbk.) | ISBN 9781315184012 (ebook)
Subjects: LCSH: Food banks--Developed countries. | Food security--Developed countries. | Poverty--Developed countries. | Corporate social responsibility--Developed countries.
Classification: LCC HV696.F6 R529 2018 | DDC 363.8/83091722--dc23
LC record available at https://lccn.loc.gov/2017052875

ISBN: 978-1-138-73973-4 (hbk)
ISBN: 978-1-138-73975-8 (pbk)
ISBN: 978-1-315-18401-2 (ebk)

Typeset in Bembo
by Taylor & Francis Books

CONTENTS

ILLUSTRATIONS

Figures

Tables

Boxes

ACRONYMS AND INITIALISMS

AD	Ahold Delhaize
APG	Agricultural Policy Note – OECD
APPG	All-Party Parliamentary Inquiry into Hunger in the United Kingdom
A2H	America's Second Harvest
BW	Bread for the World Institute, Washington, DC
BAMX	Banco de Alimentos México – Mexican Association of Food Banks
BBC	British Broadcasting Corporation
BMGF	Bill and Melissa Gates Foundation
BRC	British Retail Council
CAFB	Canadian Association of Food Banks
CBC	Canadian Broadcasting Corporation
CCHS	Canadian Community Health Survey
CEDAW	Convention on the Elimination of all Forms of Discrimination against Women
CEO	Chief Executive Officer
CESCR	Committee on Economic, Social and Cultural Rights, United Nations
CF	Carrefour Foundation
CFC	Centre for Food Policy, City University, London
CFGB	Canadian Foodgrains Bank
CFS	Committee on World Food Security, United Nations
CO	Concluding Observations
CONEVAL	National Council for the Evaluation of Social Policy, Mexico
CRC	Convention on the Rights of the Child
CRPD	Convention on the Rights of Persons with Disabilities
CSO	Civil society organization

CSR	Corporate Social Responsibility
CSI	Corporate Social Investment
CTV	CTV Television Network, Canada
DTES	Downtown Eastside Vancouver
DWP	Department for Work and Pensions – UK
EFAP	Emergency Food Assistance and Soup Kitchen – Food Bank Program, USA
EMSA	Food Security Mexican Scale
ENSANUT	National Survey for Health and Nutrition, Mexico
EPA	Environmental Protection Agency, USA
EU	European Union
EUComm	EuroCommerce
EU-SILC	European Union Statistics on Income and Living Conditions – Eurostat
FA	Feeding America
FAO	United Nations Food and Agricultural Organization – Rome
FBC	Food Banks Canada
FBLI	Food Bank Leadership Institute
FBZ	Foodbank New Zealand
FDEU	FoodDrinkEurope
FEAD	Fund for European Aid to the Most Deprived
FEBA	European Federation of Food Banks
FESBAL	Federación Española de Banco de Alimentos – Spanish Federation of Food Banks
FLW	Food Loss and Waste
FLPC	Food Law Policy Clinic, Harvard
FIAN	FIAN International for the Right to Adequate Food – Germany
FIES	Food Insecurity Experience Scale – FAO
FRC	Food Research Collaboration – UK
FS	FareShare – UK
FSC	Food Secure Canada
FSRC	Federal Surplus Relief Corporation, USA
FWRA	Food Waste Reduction Alliance, Washington, DC F90
F90	Freedom 90, Ontario
GFN	Global Foodbanking Network
GNRtF	Global Network for the Right to Food and Nutrition
HLPE	High Level Panel of Experts on Food Security and Nutrition, Committee on World Food Security, Rome
HRBA	human rights based approach
HRC	Human Rights Council, United Nations
ICESCR	International Covenant on Economic, Social and Cultural Rights
IFAD	International Fund for Agricultural Development
IWG	International Working Group – FAO
JF	JustFair – UK

LOI	List of Issues
MDGs	Millennium Development Goals
MDP	Food Distribution Programme for Europe's Most Deprived People
NDP	New Democratic Party, Canada
NFL	National Football League, USA
NGO	Non-governmental organization
NHL	National Hockey League, North America
NHS	National Health Service
NS	Nourish Scotland
OECD	Organisation for Economic Co-operation and Development
OHCHR	Office of the High Commissioner for Human Rights
OPESCR	Optional Protocol – Economic, Social and Cultural Rights
OSNPPH	Ontario Society of Nutrition Professionals in Public Health
PROOF	Food Insecurity, Policy Research – University of Toronto
QC	Queen's Council
RLOI	Reply to List of Issues
RTFC	Right to Food Coalition Australia
SDGs	Social Development Goals
SFC	Scottish Food Coalition
SNAP	Supplemental Nutrition Assistance Program, USA
SPR	State Party Report
SRRF	Special Rapporteur on the Right to Food
TT	Trussell Trust
UBC	University of British Columbia
UDHR	Universal Declaration of Human Rights
UN	United Nations
UNICEF	United Nations Children's Fund
UPR	Universal Periodic Review
USDA	United States Department of Agriculture
VCMI	Value Chain Management International, Ontario, Canada
VGs	Voluntary Guidelines on the Right to Food – FAO
VOH	Voices of Hunger
WFP	World Food Programme
WHO	World Health Organization
WRAP	Waste and Resources Action Programme – UK
WRI	World Resources Institute

PREFACE AND ACKNOWLEDGEMENTS

When contacted by Tim Hardwick at Routledge in late 2015 about writing a text on food and poverty, I was intrigued but hesitant. I had to ask myself, another book why? I had first written about domestic hunger and the import of charitable food banking from the USA into Canada in the mid 1980s. It was a time when neoliberalism directed at shrinking government and welfare states was into its stride pressed by Reaganism south of the border and Thatcherism in the UK. As food banking became further entrenched and globalized edited cross-national studies in 1997 and 2014 explored rich world hunger, charity and the right to food in separate developed nations.

In fact rather than another text I felt more like writing a polemic as indifferent governments continued to ignore domestic hunger and people's right to an adequate standard of living. As others have also written there is little evidence that food banks are an effective response to mounting poverty and increasing income inequality nor that they would fade away. In kind surplus food relief in place of fair income distribution and social security has become normalized. In face of the persistent neoliberal mantra that economic growth, job creation, financial deregulation and lower taxes would solve all problems what else was there to say.

Yet, in recognizing the perceived public legitimacy of food banking, the need to change the food charity conversation becomes daily more acute. Significantly the expanding food, public health, environmental and social policy literature on food insecurity, food waste and poverty has given little or no attention to the historical development, expansion and influence of US style food banking throughout the OECD's wealthy member states. This gap needs to be addressed.

Likewise in this gilded age of philanthrocapitalism and noting Andy Fisher's *Big Hunger* in the USA, no attention has been paid to the corporate capture and global spread to the rich world of food banking fostered by Big Food and its transnational business partners promoting food waste as a solution to domestic hunger. Over the past thirty years food charity as business has stealthily been transforming embattled welfare states and communities doing their bit into an ever expanding network of a corporately managed global food bank industry feeding the poor paving the way for publicly unaccountable food bank nations.

However, a more foundational reason for telling this story was a brief conversation with Tim Lang (professor of food policy at City University of London) at a subway station in London in late 2014. We talked about the deep hole of rich world hunger as evidence of the moral vacuum at the centre of neoliberalism and of government and public policy neglect. Certainly in writing this book, ethical considerations and the right to food as a subtext for 'joined-up' food and social policy have driven the analysis. In today's wealthy societies why are millions of people depending upon charitable and stigmatizing food handouts while being denied their human dignity? Is this the best we can do?

The day after I signed the book contract, Donald Trump became US President-elect.

Whether or not in his post-truth and continuing tax-cutting world neoliberalism is now dead, corporate food charity appears set for an even longer run and reason enough for civil society with a 'right to food' bite to develop a counter-narrative to feeding 'left over' food to 'left behind' people. No easy conversation but imperative for moving beyond food waste and surpluses to fair income distribution for all.

Food Bank Nations is about food but not about food. Principally it is written from a human rights and social policy perspective: who benefits and why from societal arrangements for advancing well-being; how much and for whom; who gives or cares for whom and why; who pays and why; and the moral choices in policy making. For their expertise advice and support in pursuing such themes over the years I would like to thank colleagues and friends Elizabeth Dowler, Tim Lang, Valerie Tarasuk, Janet Poppendieck, Tiina Silvasti, Hannah Lambie-Mumford, Sue Booth, Michael O'Brien, Leire Escajedo San-Epifanio, Karlos Pérez de Armiño, Geoff Tansey, Asbjørn Eide, Margo Young, Anne Bellows, Andy Fisher, Andrew Macleod, Adrienne Chambon and Ernie Lightman. The work of Olivier De Schutter, the former UN Special Rapporteur on the Right to Food, has likewise been decidedly influential.

There are of course many others and those who, directly and indirectly, assisted with this book. I trust they will understand that thanking all by name risks inadvertent omissions though many appear in the pages which follow. They include academics and co-authors, researchers and students, public officials, volunteers and civil society activists engaged in food policy; nutrition and public health; food waste and the environment; social policy and social action; anti-poverty campaigning; law and human rights. Their knowledge, advocacy and advice have been essential in framing the right to food discussion, and compiling the figures and tables, for which I am deeply grateful.

They also include food bank directors and staff who have assisted in charting the origins and development of food banking in OECD countries. Former and current staff in the FAO Right to Food Unit have contributed valuable knowledge as have those responsible for the development of the FAO-Food Insecurity Experience

Scale. Likewise, I have greatly appreciated the significant expertise, support and always appropriate advice offered by civil society advocates and the local, national and international NGO 'food and social justice' organizations working to ensure the human right to adequate food for all. Their analyses, commitments and actions have been particularly helpful. Any errors or misinterpretation are mine. *The Guardian*'s reporting of the food bank, food waste and food poverty issues has similarly been an invaluable source of information and analysis, as has that of *The Tyee*, British Columbia's and Canada's independent on-line magazine.

Without doubt I must thank the reviewers of the initial book proposal for their helpful ideas; Amy Johnston, Editorial Assistant, Naomi Hill, Production Editor and Elizabeth Nichols, Copy Editor for their expertise, patience and resolve for ensuring I successfully navigated the publishing guidelines and met deadlines; and especially Tim Hardwick, Commissioning Editor for his invitation and always knowledgeable, interested and valuable advice as the book took shape. It goes without saying that the continuing encouragement of my family in Canada and the UK with the space, time and support of Mary at home while writing has been essential and hugely appreciated.

Graham Riches
March 2018

FOREWORD

This is a shocking book. Shocking in its contents; shocking in that it is needed now, more than ever; and shocking to me. I have worked on food, poverty, and policy response for nearly 40 years, yet so much here comes as new or into the sharpest possible focus.

Food Bank Nations is a passionately argued and evidenced polemic against the neoliberal capture of charitable response to local experiences of poverty, manifested as people 'going hungry', in some of the richest nations on earth. Graham Riches takes on the corporate invasion and capture of what is often portrayed as ordinary people – 'good hearted folks' – trying to help out their neighbours who aren't able to feed themselves adequately by giving them food. Specific practices vary from country to country (there are examples of different histories and experiences throughout the book) but there are common threads powerfully brought together. Firstly, as a result of three decades and more of global neoliberal policies, inequality has risen inexorably. This means there are growing numbers who are too poor, through low wages, insecure work or an increasingly inadequate social security safety net, to be able to discharge their responsibilities towards themselves and their families and avoid deprivation or even destitution. Rather than construct this as a problem of poverty, by whatever definitions (and there is much debate and useful research elsewhere), the issues are framed as people being 'hungry': the answer then is to give them 'food' – they must 'be fed'. Secondly, as global institutions in the private and public sectors begin to acknowledge the crises of 'too much stuff', and particularly of 'waste', exemplified by the UN Sustainable Development Goals, there is a desperate call to reduce landfill disposal of what has been produced using increasingly precious resources of soil, water and oil – namely food – and thus to reduce food wastage and deal better with 'surplus'. Finally, the corporate sector, keen to bask in greater public approval following the exposure of greed, exploitation (of natural and human resources) and profligacy, is excellent at promoting its self-defined 'efficiencies' in management and governance, streamlined delivery systems (including IT and transport) and systematic use of human and other resources for profit.

Put these three together – (food) poverty, waste, streamlined systems – and the idea that the corporate sector, with immense skill, efficiency and generosity, channels food surpluses towards those who are 'in need', can be seen as a 'no-brainer'. It seems an obvious solution to several problems; thus, most citizens are encouraged not to concern themselves with either why the problems occur, nor what might happen if such charity did not exist. Riches' analysis shows, by contrast, that what is happening is precisely designed to discourage critical thinking and appears as a simple, linear solution when in fact we all face a complex, challenging set of problems to which real solutions might be more uncomfortable. At a time when tax avoidance and poor working conditions, along with rising inequality (Alvaredo et al, 2018) are very much in the news, alternative solutions to charitable food ought increasingly to be in the public mind too. OECD nations are not used to food riots, or people marching for decent jobs and wages, or widespread strikes in support of better work and pensions. They do not, on the whole, welcome or support reminders that all citizens – indeed, all people – have fundamental needs of subsistence, protection, affection, understanding, participation etc (see Max-Neef, 1991; also ideas developed by Raworth, 2017) which should be met in ways which do not undermine human dignity. Using charitable systems to mitigate immediate need arguably dampens down any thinking along such lines. Furthermore, 'charity' is big business and mostly tax-deductible; it is also largely unaccountable to those who receive its beneficence. Thus 'food banking nations', as is so powerfully set out here by Riches, have become the epitome of neoliberal thinking, in that the corporate sector is permitted both to appear to solve two of the problems it helps create (waste or surplus, and low wages or insecure employment) and to divert attention from the social and political challenges they pose. Finally, charity undermines the role of government in addressing its responsibilities for wellbeing, fulfilling the neoliberal agenda of 'rolling back the state' and its interference. No wonder that corporate involvement in food charity is so widespread, powerful and largely hidden from view.

The joy of Riches' writing, here and elsewhere, lies in his bringing all this to light. He has a deep understanding of how these issues have interacted over many years and in different places, an understanding which comes from having been involved in practice and research, as well as powerful writing, when it was not fashionable to do so and now when it most certainly is. Secondly, his belief in rights-based approaches to matters of social policy have profoundly informed his thinking, to the benefit of colleagues and, more importantly, those living in tough circumstances and/or trying to address them with practical and policy responses. For instance, in the UK, the Manchester based charity *Church Action on Poverty*, long experienced in enabling the voice and agency of those who are poor, including food bank clients, to be heard in debates about and responses to poverty, has espoused the right to food with energy and commitment. Similarly, *Nourish Scotland* is drawing on rights-based approaches in its advocacy of policy change.

I have personal cause to be grateful to Graham Riches for his insights and encouragement over many years. After working in the global South on malnutrition and poverty, and helping build capacity to use nutritional conditions and indicators for advocacy to effect policy change, in the early '90s I shifted to help bring such analysis to my own country of the UK. With others, I argued that poor nutritional outcomes and limited food consumption are not due to inadequate parenting or budgeting skills but are markers of a dysfunctional food and economic systems, in which people are systematically impoverished by economic strategies which bypass the needs of the many in order to strengthen the power, and wealth gain, of the few. Policy response should strengthen and support the resilience strategies people themselves adopt or prefer, rather than undermining them. As in the global South, social and economic inequality and poverty are critical in shaping food access and practices, while policies from the state and the private sectors have tended to make things worse. In the early '90s (as now) local initiatives to help with cooking skills, were the standard – and inadequate – policy response. The resonance with much work in the global South was striking, as was the fact that people with few material resources nevertheless cared deeply about what they ate. As a nutritionist I had had to learn that a diet is more than a bundle of key nutrients: food represents status, pleasure, identity – offering food is a fundamental aspect of hospitality, and shows both who you are and who you respect and love. This is as true for those who are poor as for the richer, and has to inform policy.

How disturbing then, in the early 1990s when the UK was espousing neoliberalism, to find that the much-vaunted social security system was completely inadequate to provide sufficient money for people to live on for more than a few months with any kind of decency; that school meals, including those free for children of some social security claimants, no longer offered what a child needed for healthy growth and development (provision having been deregulated by the British Government, both in terms of nutritional standards, and who could offer it). Arguments put forward about entitlement to and levels of social security, from civil society, rights workers and a few academics, were dismissed by Government, with the support of media that carried popular opinion: despite growing evidence to the contrary, social security levels were deemed sufficient if people budgeted properly. Free food, through soup kitchens or drop-in centres, was offered to those with no other means of obtaining food (essentially those described as homeless) and community rights workers or Citizens Advice Bureau staff would occasionally mention the small stash of cheap tins and packages of basic foodstuffs they kept to hand out, surreptitiously, to those sometimes left without money through administrative error. I knew from colleagues in Australia, New Zealand, France, Germany and Portugal that similar conditions and circumstances were emerging. These experiences of food poverty in the global North were discussed at two international European conferences and in subsequent books (Köhler et al, 1997; 1999), but 'food banks' were not then centre stage. In Canada and the USA, by contrast, such charitable food was gaining ground, as Riches so powerfully documents. In the mid-1990s I contributed to his earlier book on rich countries

(Riches, 1997) surveying and contrasting experiences of poverty and hunger, and different policy responses.

Roll forward two decades and a hitherto unimaginable story is now told, in the UK as in many OECD countries. What was marginal practice and only for the near destitute – giving out free food donated by concerned citizens or supermarket 'waste' – is now standard practice to meet many citizens' needs, and has become in many senses very big business. It is a practice recognized and accepted, indeed welcomed, by the general population and by government in many rich countries. The fact that the general approach of giving surplus food as aid has long been shown to be an inappropriate and inefficient response to poverty and poor nutrition in the global South is totally ignored. The fact that doing so contravenes and undermines people's rights to access food appropriate for healthy living by purchase or household production, is dismissed as irrelevant. The fact that large corporations, themselves part of why waste and poor working conditions are so widespread, receive social, political and economic gain from these new practices, is unremarked upon. There is resistance and protest against this corporatization of human misery which are documented in this important book, not least in civil society and among concerned citizens, whose understanding and espousal of collective solidarity is clear and strong. Such sites of resistance need strengthening and shouting out – they are the true and only way forward, and Graham Riches is to be heartily commended for having set out so clearly the pernicious nature and moral illegitimacy of food banking nations, and the critical possibilities and hopes of doing things differently.

Elizabeth Dowler
Emeritus Professor of Food and Social Policy
University of Warwick, UK

References

Alvaredo, F., Chancel, L., Piketty, T., Saez, E. and Zucman, G. (2018) *World Inequality Report*. World Inequality Lab http://wir2018.wid.world/

Köhler, B., Feichtinger, E., Barlösius, E. and Dowler, E. (eds) (1997) *Poverty and Food in Welfare Societies*. Sigma Edition

Köhler, B., Feichtinger, E., Winkler, G. and Dowler, E. (eds) (1999) *Public Health and Nutrition*. Sigma Edition

Max-Neef, M.A. (1991) *Human Scale Development Conception: Application and Further Reflections*. Apex Press

Raworth, K. (2017) *Doughnut Economics: Seven Ways to Think Like a 21st Century Economist*. Penguin Random House.

Riches, G. (ed.) (1997) *First World Hunger: Food Security and Welfare Politics*. Macmillan

1

INTRODUCTION

Wasted food for hungry people

In the world's most affluent and food secure societies it is now publicly acceptable for corporate food charity to feed wasted food, surplus to the requirements of a dysfunctional industrial food system, to millions of hungry citizens who are surplus to the requirements of the labour market. Surely scandalous food waste cannot be the solution to the injustice of widespread domestic hunger in the rich world, or anywhere else.

As eminent French historian Fernand Braudel commented in 1985 when asked about the gap between the 'new poor' and the rich, coincidently at the very time when food banks were first established in France, he remarked that 'today's society, unlike yesterday's, is capable of feeding its poor. To do otherwise is an error of government', also noting that 'France is stingy with its wealth' (Braudel, 1985). In the more than three decades since, fellow rich world OECD governments have largely remained indifferent to his words confident there is no alternative to market driven neoliberalism and austerity and punitive welfare reform whenever deemed necessary. Little wonder that surplus food for hungry people has become normalized and publicly accepted. Yet there are always choices and alternative courses of action: moral, lawful, political and practical.

Adopting a human rights based approach, *Food Bank Nations* disputes both the effectiveness and moral legitimacy of this publicly unaccountable food banking and casualty treatment model of feeding hungry people in OECD member states. It is a challenging story for the many who organize and operate charitable food banks and especially those who support them as volunteers giving their time, expertise, donations and most of all their human compassion to feed hungry people. Such public spiritedness surely fits well with the 2015 UN Social Development Goals which plan to eradicate global hunger and cut in half per capita food waste at retail and consumer levels by 2030 (SDGs, 2015a). Yet in the world's wealthy nations is the food bank contribution the best we can do?

Uncomfortable issues and pointed questions are raised regarding our business and political elites who leave domestic hunger to charitable good will. It is definitely not a story about being fed nor of building bigger and better food banks and food safety nets. It is about the universal right of vulnerable individuals and families to be able to feed themselves with choice and human dignity. It presses the case for rethinking what we mean by solidarity with the poor and why the human right to adequate food matters to us all.

The book traces the global spread of charitable food banking from its origins in the USA in the late 1960s to its entrenchment in today's OECD nations club. Along the way it explores the hidden functions of Big Food's corporate capture of charitable food banking in the rich world which allows us to believe that food waste and domestic hunger are being effectively addressed. This permits indifferent governments to look the other way with public policy neglect ensuring food poverty is left untouched. It asks why the main beneficiaries of food banks are the tax incentivized food retail sector with reduced landfill costs and the branded corporate social responsibility seal of approval; and austerity minded governments bent on fiscal restraint, social spending cuts and regressive taxation benefiting the well fed rich and powerful. Meanwhile the task of holding the State to account falls to civil society.

Searching for answers to domestic hunger food banking is a powerful metaphor for the deep hole at the centre of neo-liberalism, and of the moral vacuum and lack of public accountability at the centre of government decision making. It is a story of 'left-over' food for 'left behind' people enduring the pain of stigma and the loss of human dignity; the advance of the corporately dependent charity economy; the abandonment of social rights and neglect of public policy; and the glaring absence of political will.

Food Bank Nations is about the false promises of economic growth, job creation and employment (Forrester, 1999; Moore, 2016); the plight of dispossessed people condemned to begging and wasted lives (Bauman, 2004); the re-emergence of absolute poverty and widening income inequalities; broken social safety nets and the retreat to long-term emergency food aid; the shift from income assistance to food assistance; the undermining of the welfare state and the re-writing of the social contract (Ha-Joon Chang, 2016). The book endorses the view that while 'the dominant neoliberal paradigms of today consistently claim that social policy is less important than economic policy', that 'to the contrary, conceptually, economic policy is but a small subset of a broader social policy and that the evaluation of economic initiatives should use the same normative criteria that are applied to traditional social programs' (Lightman and Lightman, 2017).

Whilst recognizing the moral imperative to feed hungry people *Food Bank Nations* challenges the effectiveness, sustainability and moral legitimacy of globally entrenched corporate food banking as the primary response to rich world food poverty. Thinking and acting outside the food waste and charitable food aid box requires changing the public conversation from charity to one grounded in social justice, human rights and why the right to food matters. The subtext of the right to food is public policy as an expression of *critical* and *collective* solidarity which ensures food security and food justice for all. The UN Food and Agricultural Organization (FAO) rightly preaches the efficacy of the right to food and good governance to the Global South (FAO-VGs, 2005). It is a powerful message which urgently needs to be heard in the prosperous Global North.

Why food matters

In June 2015 I was a member of a panel to discuss the findings of a just completed report entitled '*Hungry for Justice: Advancing a Right to Food for Children in BC*'. It was held in Vancouver's Downtown Eastside (DTES), Canada's poorest postal code only too well known for the seemingly implacable range of economic, health and social issues which confront it. It is a community with an immensely proud reputation for its inner resilience and locally led anti-poverty activism and is the site of numerous health and social agencies and emergency feeding programmes. Participants at the event included long time DTES residents and food and community activists both from within the neighbourhood and the city.

In her introduction Laura Track, a lawyer then with the British Columbia Civil Liberties Association and author of the report remarked that she herself had 'never truly experienced real hunger'. It was a respectful acknowledgement that hunger existed outside her personal experience, as indeed my own. In my view it was a political statement recognizing the challenge of food inequalities and injustice. It was especially a reminder that food, and access to it, should never be taken for granted, as an everyday and unexamined entitlement for those on the wealthy side of the tracks. When thinking and acting about hunger in the rich world whether as academics or professional experts, anti-hunger activists, food bank volunteers, members of the public and especially politicians we first need to remind ourselves as to why food matters in society and in our own daily lives. We need to grasp a little of what it means to be without food, to be hungry.

It should not need saying that food and water are basic human needs and fundamental to diet and nutrition, to human growth and development, to physical and mental health and to life itself. However, to the extent that governments of the world's rich nations ignore the millions who cannot afford to eat, nutritiously or not, the moral and political nature of this denial and indifference needs continuously exposing until action is taken.

Food as a political matter

Food clearly has many dimensions and matters for many different reasons. Writing from a concern that people need access to healthy food choices Aileen Robertson and colleagues, reveal the key aspects and the complexity of the food system while pointing out that food is primarily a political issue (see Box 1.1).

BOX 1.1 FOOD IS ABOUT

Politics: because global trends may threaten food security and traditional culture.

Economics: because food production, processing and marketing contribute about 30% of gross domestic product in most countries.

Socio-economic change: because of urbanization and movement away from agricultural employment; these and other factors, affect the availability of and access to employment and consumption of food.

Environment: because food is grown, transported, stored, processed, packaged and wasted. Sustainability from the farm to the kitchen could save much energy.

Science and technology: because new developments are changing the food supply. Genetically modified foods, precooked meals, dietary supplement and zero-calorie fats challenge traditional values about diet.

Health: There is a global epidemic of diet-related non-communicable disease, while at the same time undernutrition afflicts some 1 billion of the world's population.

Source: Robertson et al, 1999, p.180

Importantly the authors comment that 'the food chain from plough to plate needs scrutiny to ensure that healthy food choices are the easy ones and unhealthy choices more difficult' (Robertson et al, 1999, p.180). These are highly significant policy matters for governments and political decision making and ones which in the first instance must ensure that vulnerable populations have access to food in the market place. In that context we need reminding, as professor Tony Winson points out, that food is a commodity with 'its own very specific and significant socio-economic, cultural and political characteristics'. As a political commodity, food is fought over within and between national governments and transnational Agri-Food and Big Food corporate interests and nation states (Winson, 1993).

Politically food production and supply is central to feeding the world and as such to global and national debates about food policy within and between rich and poor countries alike raising issues and consequences for agriculture, the environment, health and trade. By the same token food distribution and access to food in the Global South and North are matters of social policy and income distribution when addressing domestic hunger, poverty and inequality. Therefore who controls and shapes the food system and who has access to it are matters of concern to society as a whole. If we believe that food belongs to us all then we need to be alert to the power of the transnational Big Food conglomerates upon whom today's charitable food banks are very much dependent. After all Big Food's bottom line is safeguarding the food system in the interests of its shareholders before those of the public interest. Is what's good for business good for us all, and the poor's access to healthy food choices?

Food as an 'intimate' commodity

Anthony Winson who writes about the political and economic determinants of diet and nutrition also notes that 'within capitalist societies food has become a commodity – one in a long line of goods and services produced and sold' (ibid). As an economic commodity many livelihoods depend on the agro-food industry for its production, processing and retail. Profoundly he also makes the point that 'food and water are "intimate" commodities like no other. In the process of their consumption we take them inside our very bodies, a fact that gives them a special significance denied such externally consumed commodities such as refrigerators, automobiles....' (ibid).

Not only is food essential for personal health it is a social and cultural good vital to our sense of individual, family and social well-being. Food has symbolic meanings and is deeply embedded in different cultures and world religions and is central to family and community life. The idea of companionship derives from the ritual and practice of taking bread together. Food can be about pleasure, good company or romance. As such it reaches into every corner of human existence and social experience. In this sense food belongs to us all. It is the connective tissue between the individual and community and acts as a powerful force for integration within and between societies, rich and poor. Ensuring the human right to adequate food for all is an expression of collective solidarity.

Food deprivation as social exclusion and injustice

Food is also an essential commodity, as Winson writes, 'we literally cannot live without it' (ibid, p.4). Being carless, he suggests, may be a major inconvenience but 'to be deprived of food is nothing less than "catastrophic"' (ibid). When governments deprive people of food for those whose welfare benefits have run out; or who have been punitively sanctioned; or whose wages are insufficient to put food on the table; or as in Indigenous communities in Australia, Canada, New Zealand and the USA – whose land has long since been appropriated – they are not only infringing their rights of citizenship but are denying access to their full participation in society; and to life itself. In these ways the State renders them 'other' and thereby manufactures injustice.

Of course if following Margaret Thatcher's neoliberal dictum 'there is no such thing as society', being unable to feed oneself and one's children would be no concern to government. Rather the responsibility would rest entirely on the hungry and homeless themselves or left to poor law food relief as in Victorian times or corporately dependent charitable food banks as today.

The late internationally respected Peter Townsend in his classic UK study of household resources and living standards in the 1960s well expressed this sense of social exclusion and injustice, and of relative deprivation. People, he wrote, were in poverty when they lacked 'the resources to obtain the types of diet, participate in the activities, and have the living conditions and amenities which are customary, or at least widely encouraged or approved, in the societies to which they belong. Their resources are so seriously below those commanded by the average individual or family that they are, in effect, excluded from ordinary patterns, customs and activities' (Townsend, 1979).

The inability of people to access an adequate diet formed a basic indicator of deprivation including such items as clothing, fuel, housing, security at work, family support, education and social relations. Townsend's research showed there were income levels below which it was not possible to participate in normal and customary activities which became a key measure of relative deprivation. Social solidarity was denied.

In this context I recall a class at UBC discussing the procedures by which social workers on a home visit might conduct child welfare risk assessments. The fridge is one item for inspection. What does it mean if it is empty or just containing a few scraps of processed food? Is it simply checked off the list or is this the moment to explore the issues confronting the family? What is the human story behind the lack of food? What are its implications for the family which is hungry, for social work practice and for social policy? How to ensure their human right to adequate food?

This is not an isolated example. Such material deprivation is widespread and still acutely felt. Doubtless it is experienced more acutely today in most if not all OECD countries as income and wealth inequalities have widened. In 2012, Tim Lang, leading expert on global food policy at City University in London, speaking at an OECD Conference on 'Mobilising the Food Chain for Health' reminded his audience that the UN Special Rapporteur on the Right to Food, Olivier De Schutter wished to draw policy-makers' attention to the realisation 'that food is a significant source and indicator of social inequalities and injustice' (Lang, 2012 p.2; De Schutter, 2011).

The problem is food has become so everyday (Visser, 2000). It is taken for granted it. Let alone being unable to feed themselves, millions of families in high income and 'food secure' countries are unable to invite their neighbours in for a meal, a cup of coffee or tea or to celebrate high days and holidays without being embarrassed because of lack of income (Fabian Society, 2015) and are unable to shop for food on a regular basis.

Indeed being denied access to food and having to cope with bare fridges and empty plates highlights significant issues of personal and social identity. As Suzi Leather has said 'food is of course much more than a vehicle for nutrients. Food choice is an expression of the belonging aspects of life: family habits, regional identity, national, religions and cultural tradition' (1996). Elizabeth Dowler, emerita professor of food and social policy underlines this point: 'in the general public's mind' food 'represents an expression of who a person is, where they belong and what they are worth, and is a focus for social exchange' (Dowler, 2003). It is essential for fully participating in society. Food touches everything. It matters to us all in rich and poor worlds alike.

Health and financial costs to society

As for dietary and nutritional well-being we likely are what we eat (Pretty, 2004). Yet, by the same token, being deprived of food has many long-term consequences for individual health and well-being with huge financial costs for society. As Carolyn Shimmin and Valerie Tarasuk, professor and director of the PROOF food security research centre at the University of Toronto have noted, for children food insecurity 'is associated with poorer physical and mental health outcomes, including the development of a variety of long-term chronic health conditions such as asthma'; and for adults is 'independently associated with increased nutritional vulnerability, poor self-rated health, poor mental, physical and oral health and multiple chronic health conditions including diabetes, hypertension, heart disease, depression, epilepsy and fibromyalgia' (Shimmin and Tarasuk, 2015). The lack of food due to financial constraints matters as much as the food we do eat.

As Tarasuk and colleagues through their pioneering research have also pointed out the financial costs to the Ontario provincially funded health care system are significant. In Canadian dollars a food secure individual costs on average $1608 p.a, compared to $2806 for the moderately food insecure, and almost $4000 for severely food insecure individuals, those who are going hungry (Tarasuk, et al, 2015). South of the border in the USA, which lacks a universal health care system, the costs are even more eye opening. According to a recent report by the Christian anti-hunger Bread for the World Institute, 'America's hunger health costs tally up to a staggering $160 billion annually'... 'affecting educational outcomes, labour productivity, crimes rates, Gross Domestic Product, and much more' (Rotondaro, 2015). As for public policy, domestic hunger and food insecurity clearly should matter.

Being deprived of food primarily results from having insufficient money in your purse or your pocket, or from your inability to use your debit or credit card to go into a store like anyone else and purchase the food of your choice. Most tellingly it is a marker of being disconnected, excluded from and rendered surplus to mainstream society and what is normal and customary. It is about being disempowered and being rendered other; being held to ransom by punitive welfare legislation or inadequate wages; suffering the shame of begging from charitable food banks and being denied your human rights.

Tony Winson succinctly makes the case for why food matters for each of us as individuals. He quotes a student of the Pueblo Indians of the south-western United States who said 'when I take away corn from such people, I take away not only nutrition, not just a loved food. I take away an entire life and the meaning of life' (quoted in Farb and Armelagos, 1980, p.7). If we wish to have a constructive dialogue about food charity we first need to understand what it means to be hungry or food insecure in the prosperous and grossly unequal Global North.

Food banking, early days

My attention was first drawn to food banks when teaching social policy in western Canada in the early 1980s. One had just been established in Regina the provincial capital of Saskatchewan. Having no idea what a food bank was I quickly learned they were centralized warehouses registered as non-profit agencies which collected, stored and distributed surplus food (free of charge) to charitable agencies or directly to hungry people themselves. The first food bank was set up in Phoenix, Arizona (1967) as a response to the 1960s rediscovery of poverty and hunger in the USA. In 1979 Feeding America was established, the national clearing house for food banks across the country.

In the early 1980s the US food banking model of providing emergency food assistance was imported across the border into Canada (Riches, 1986). I was as shocked and curious then as I remain today by this unexpected arrival of direct food relief and charitable food banking. I should not have been so shocked. These were the early days as the neo-conservatism preached by the UK's Margaret Thatcher (1979), Ronald Reagan (1981) in the USA and Brian Mulroney (1984) in Canada who dug the foundations of the minimalist state.

Nevertheless why were food banks necessary in Canada? After all the Salvation Army and St Vincent de Paul were long established charitable missions providing emergency relief; Canada's welfare state provided a comprehensive system of social security including income assistance and a publicly funded social safety net; and Canada is one of the world's high income states with membership of the OECD and the G7, the world's rich nation clubs. In aggregate terms through local food production and imports we live in a food secure country; and moreover the agricultural prairie provinces in western Canada are part of the global bread basket exporting grains to feed a hungry world.

Notably and ironically, in 1976, five years before the establishment of its first food bank, Canada had ratified the *International Covenant on Economic, Social and Cultural Rights* (*ICESCR*, 1966, 1976) including the right to food as a key element of the right to an adequate standard of living. Today the *ICESCR* has been ratified by 164 member states but not, it should be noted, by the USA.

Yet despite the affluence of Canadian society, its overabundance of food and formal recognition of its obligations under international law to 'protect, respect and fulfill' the right to food, Canada has chosen not to ensure the food security of its own vulnerable populations including most notably its Aboriginal and First Nations peoples, and those leading precarious lives outside or on the edge of the labour market. Disturbingly, as later research and *Food Bank Nations* reveal this indifference (some would say refusal) to safeguard the right to food of its poorest citizens is not a uniquely Canadian story. It is one being told in many rich nation states (Riches, 1997; Poppendieck, 1998; Dowler and Jones Finer, 2003; Riches and Silvasti, 2014; Selke, 2015; Caraher and Coveney, 2016; Fisher, 2017)

The establishment of the Regina Food Bank in 1983 was not an easy decision. As Eldon Anderson, the first board chair, explained at the time there were strongly held concerns about the dilemmas in resorting to food charity to feed hungry people. Eldon had been born in a three-room homestead farmhouse and raised as a child in the dirty thirties (Leader-Post 2014). He had clear memories of the breadlines of the Great Depression. He also recalled the period following the end of World War II when Canada was alive with the promise of building a new society and a home fit for heroes in which the breadlines of the 1930s would be a distant memory. During the post war period as Canada's national economy and prosperity rose and its welfare state, influenced by Beveridge's ideas in the UK, slowly developed a robust system of income security, social services and publicly funded health care (similar to the NHS) was put in place (Guest, 1997). Eldon had never imagined he would once again see his fellow citizens standing in the breadlines begging for parcels of food.

This however was the new reality, a consequence of a major economic recession and significant joblessness in the early 1980s with income-based unemployment insurance and social assistance (welfare) programmes failing in their legislated mandates to ensure adequate benefits. Food banks began to pick up the pieces of a failing social safety net and become the early warning signs, symptoms and symbols, of the retreat from the welfare state. The mantra of neo-liberalism hastened the withdrawal as deregulation, privatization and the politics of the minimalist state gathered pace.

Critical of such public policy neglect and despite unhappy memories of soup kitchens and charitable food relief during the Great Depression, Eldon was committed to organizing the Regina Food Bank in 1983. As his daughter Tanya Humphries writes he passionately believed that 'we are our brothers keeper' and in helping people in poverty 'we have no idea what they have been through or what they are battling ... go give him a smile and this bill', usually a $100 (TH – email communication, 7.7.17). In other words practical compassion dictated the need to act immediately, not waiting for tomorrow. Emergency food assistance would hopefully be just a short-term response, though as we now know a hope long since dashed in an increasingly wealthy and unequal first world.

As the rise of the food bank nations shows market driven global philanthropy has gone from strength to strength in the process becoming institutionalized and corporatized all the while allowing unaccountable governments to avoid their human rights obligations to address domestic hunger. Regrettably necessary as many would argue charitable food banks are compassionate and practical responses to domestic hunger and food poverty in today's rich, food secure world. They are the logical outcomes of thirty years and more of neo-liberalism, the false promises of economic growth and of public policy neglect. Heavily dependent on Big Food for the sourcing of surplus and wasted food to feed hungry people, they are reliant on the free labour and tireless giving of volunteers doing their bit to take the bite out of poverty and austerity.

Food bank nations, today

In helping to put food on the table, food banks argue they act in solidarity with millions of marginalized people in wealthy countries unable to feed themselves and their families. In North America and for the most part in Europe they have become like food itself, every day and taken for granted. In the UK where the debate about their role is somewhat sharper, they are nevertheless regarded as expressions of former Prime Minister David Cameron's Big Society and likely exemplars of 'effective altruism' (MacAskill, 2015). Certainly, preaching solidarity, the moral imperative informing their work is unimpeachable but one must ask, solidarity with whom?

As Martha McMahon, sociologist at the University of Victoria in British Columbia has observed 'the new market focused philanthropy doesn't get the kind of scrutiny it needs – how can anyone question someone who wants to end poverty or hunger', astutely commenting that the 'the ideology of philanthropy inhibits serious critique' (McMahon, 2011; see also McGoey, 2015). Precisely so. It therefore has to be asked whether food banks really are the best that wealthy societies can do in addressing widespread food poverty. Morally and effectively what form of solidarity are they expressing?

In this current Gilded Age dominated by years of neo-liberalism in the OECD world, should we not be asking whether 21st-century food banking – today's re-invention of 1930s food relief and soup kitchens – is better understood as a disabling force disempowering the hungry and those who argue for food and social justice for all. Rather than expressing social solidarity, food banking more likely functions as a form of 'uncritical' solidarity (Pérez de Armiño, 2014) serving as a daily reminder of the empty rhetoric of the human right to adequate food and undermining public policy. Who really benefits and why from the free distribution of corporately 'donated' surplus food? Given the shift from income assistance to long-term emergency food – in other words food safety nets – are food banks more a hindrance than a help in addressing food poverty?

The corporately tilled soil of surplus and wasted food (Stuart, 2009; Bloom, 2010; HLPE, 2014) and the neo-liberal roots which continue to nourish the growth and development of the food bank nation tell the story that charitable food relief – food aid, food assistance, food parcels, food handouts – in masquerading as a social safety net is very much part of the problem and not the solution to pervasive food poverty in the rich world. The myth of food charity (Poppendieck, 1998; Berg, 2008) more likely obscures the unpalatable fact that the blight of hunger in today's rich food secure world is a direct consequence and continuing moral indictment of thirty years plus of political amnesia and public policy neglect. Governments, deliberately or not, have decided that food charity is the preferred response to domestic hunger.

It should therefore come as no surprise that concepts such as food security and insecurity, malnutrition and food poverty, food aid and emergency food assistance are no longer restricted to addressing hunger in the Global South but increasingly inform food and nutrition, public health, social policy, environmental debates and media dialogue about hunger in the prosperous Global North. Food banking perceived as the strategic link between food waste and food poverty reduction joins this lexicon as an effective 'win-win' solution to both issues (GFN, 2011; FBLI, 2016).

Let's face it, in the early years of the 21st century charitable food banks in the first world are unremarkable, normal and publicly acceptable frontline community agencies doing their bit to deliver emergency food assistance – wasted and surplus food – to hungry people. In the USA you would need to be a baby boomer to recall a time without food banks and in Australia, Canada, and New Zealand born before the early seventies. UK millennials may however just remember a time before food banks as their arrival is a more recent phenomenon largely driven by the Great Recession of 2008, austerity budgets and punitive welfare sanctions (APPG, 2014). Their expansion has been fast tracked. The Trussell Trust has even proclaimed that every town should have one (Lambie-Mumford, 2013), a clear call supporting the rich world's food bank nation club.

Within the OECD food banking is evident across Europe and around the globe. The European Federation of Food Banks (FEBA), dating from 1987 and centred in Paris, is active in 27 (EU) countries either with national food bank federations (23), single food bank organizations (4) or supporting food bank projects (FEBA, 2016a). It also has a partnership agreement with Die Tafel, the food banking network in Germany. FEBA works closely with the Chicago based Global Foodbanking Network (GFN) established in 2006. The GFN is active in 32 countries around the world within and outside the OECD including high and middle income countries including Australia, Brazil, Canada, Chile, China, Hong Kong SAR, India, Israel, Mexico, Russia, Singapore, South Korea, Taiwan, the UK and the USA (GFN, 2016a).

The endorsement of Big Food's transnational corporations has been instrumental in food banking's global spread and its entrenchment in OECD member states. GFN's current corporate partners include US food and drink conglomerates such as Cargill Inc and Kelloggs (founding partners), General Mills, H-E-B and Unilever (GFN, 2017a) with Coca-Cola, Conagra, Pepsico, Costco, Mondolez (formerly Kraft) and Walmart as significant supporters. In Europe FEBA has likewise benefited from Cargill and Unilever's significant support as has food banking from other retail, food services and processing giants such as Asda (UK); Carrefour and Sodexo (France and Spain); Unilever (Netherlands/UK) and Nestlé (Switzerland).

As with the transnational spread of processed foods and the US fast food industry depicted in Eric Schlosser's *Fast Food Nation* (2001; 2012), corporate food charity has likewise built itself a long and enduring shelf life. Food banks, like fast food restaurants are now entrenched in the global food supply chain. Both are products of a wasteful industrial food system. Yet despite cheap food and free handouts of surplus food millions of people in the rich world still cannot afford to put food on the table. More to the point fully stocking food bank shelves rarely matches demand.

The right to food

In its critique of the corporately captured charitable food aid box, the book's central thesis, informed by food and social policy analysis, focuses on food as a human right seen through the lens of collective solidarity: the moral, legal and political obligations of the State to ensure food access for all. While food is produced, sold and bought as a commodity within a global market economy, it is also understood as a public good, part of the commons and belonging to us all. *Food Bank Nations* explores food poverty and domestic hunger in the rich world as issues of individual and collective human rights and profoundly political matters.

The right to food, in the words of the UN Special Rapporteur, is

> the right to have regular, permanent and unrestricted access, either directly or by means of financial purchases, to quantitatively and qualitatively adequate and sufficient food corresponding to the cultural traditions of the people to which the consumer belongs, and which ensure a physical and mental, individual and collective, fulfilling and dignified life free of fear.
>
> *(OHCHR, 2017a; Ziegler et al, 2011)*

The right to food is not about charity and being fed but about the right of all to feed themselves with choice and human dignity. I have not substituted the accepted RTF acronym for the right to food choosing instead to write it as commonly spoken and referring to the human right to adequate food and nutrition.

Book outline

The human right to adequate food frames the discussion regarding the rise of *Food Bank Nations* and what to do about the corporate capture of domestic hunger and food charity in wealthy OECD nation states. It asks what do we mean by hunger and how to explain the persistence of first world food insecurity? What if anything has surplus and wasted food got to do with solving the food poverty issue? How acceptable or concerning is the long reach of the US inspired food bank nation? What are the hidden functions of corporate food charity and how effective is the food bank benefit chain? Who really benefits and why from diverting increasing volumes of food waste from landfills to food banks. Is food bank nation the path to follow or are there alternative courses of action: public policy and the right to food? Does the right to food matter, and if so what is the role to be played by civil society. Such questions are explored in the chapters which follow which pursue four overall themes: *domestic hunger and charitable food banking; corporate capture; rights talk and collective solidarity; and changing the conversation to the right to food.*

Domestic Hunger and Charitable Food Banking: considers the nature, prevalence and causes of food poverty and hunger in the rich world (Ch. 2) setting the context for examining the rise, national institutionalization and spread of charitable food banking in OECD member states. and growing dependence on surplus food. It traces the rise of food banks from their local origins in the USA in 1967, following their post 1981 journey crossing OECD national borders and appearance in Sweden in 2015 while considering different food banking models (Ch. 3).

The following four chapters focus on the *Corporate Capture* and the global consolidation of food banking by Big Food and its thriving corporate and philanthropic partnerships with Feeding America, the European Federation of Food Banks, the Global Foodbanking Network and the Food Bank Leadership Institute to the growth spread of food safety nets (Ch. 4). While food waste and food poverty are significant but separate issues, the question posed is whether diverting wasted food from landfills to food banks in the process manufacturing surplus food can be a solution to either. Who benefits and why (Ch. 5), leads to the question whether corporate food banking is part of the solution to hunger (Ch. 6) or the problem advancing false promises of 'uncritical' solidarity allowing neoliberal driven and indifferent governments to look the other way (Ch. 7).

Rights Talk and Public Policy reframes the issue of domestic hunger in the rich world as matters for human rights and the State. Informed by collective solidarity and the principle we are all 'rights holders', it discusses the right to food in international law and the moral, legal and political obligations of the State to act as the 'primary duty bearer' ensuring food access for all (Ch. 8). It considers the work of UN institutions in the mainstreaming of human rights and holding OECD food bank nations and indifferent governments publicly to account through international monitoring of their compliance with the right to food (Ch. 9). Attention is drawn to what the right to food is (and is not), to its implementation and the crucial role of civil society with a strong right to food bite for reclaiming public policy (Ch. 10).

Gathering Political Will focuses on changing the conversation towards a right to food agenda (Ch.11) presenting challenging propositions for public debate. Recognizing the moral imperative to feed hungry people, it reviews why corporate food charity blocks solutions to rich world domestic hunger; and why publicly accountable 'joined-up' policies demand the State's attention. In asking whether feeding surplus food to hungry people really is the best we can do, it poses questions for civil society and public debate informed by collective solidarity and putting human rights and politics back into hunger. It seeks civil society with a right to food bite. Food bank volunteers are doing their bit. Will the Indifferent State show up? Will politicians do theirs?

PART I

Domestic hunger to charitable food banking

2

FOOD POVERTY AND RICH WORLD HUNGER

It may seem improper, even unthinking to speak of hunger in today's affluent nation states when as Elizabeth Dowler has asked 'how can the extent and depth of poverty, hunger and nutritional inadequacy in rich, industrialized countries be in any way compared to the experience of those in the "South"' (Dowler, 2003), when in the 'North' as she notes, anti-poverty strategies and welfare states have been in existence for at least a hundred years. How can anyone living in the OECD wealthy members club be hungry and unable to feed themselves and their families.

After all the Organisation for Economic Co-operation and Development was founded in 1960 to stimulate economic progress and world trade. Comprising 33 high and two upper middle income member states with a population of 1.3 billion, the OECD is committed to democracy and the market economy. Its mission is to promote policies that will improve the economic and social well-being of people around the world (OECD, 2017a).

Yet today all OECD member states have become food bank nations. Twenty-one national food bank associations are members or affiliated with the European Federation of Food Banks (FEBA); 8 belong to the Global Foodbanking Network (GFN); and 6 have food banks in various stages of development. Charitable food banks are markers of domestic hunger with national and international surveys telling us that food poverty is widespread in this rich world members' club.

Thinking about food poverty and access to food

The meanings we attach to food poverty, food insecurity and domestic hunger matter. Who is monitoring it; how it is measured; who is at risk are central questions, especially food access and affordability. The problem is that many politicians in wealthy states dismiss the idea of domestic hunger.

Denial back then

In the early days of food banking in British Columbia, Grace MacCarthy, the Minister of Human Resources and responsible for social assistance declared on national television that compared to the developing world there was no such thing as hunger in the province: no one was starving to death (CBC, 1985). Yet, I recall Ed Bloos, the director of the Regina Food Bank always rebutting the criticism that people who used food banks were not hungry. He would ask why would people stand in the freezing cold, in wintertime Saskatchewan the temperature dips to −30 and lower, to obtain a bag of sub-quality processed foods if they were not hungry and did not know where the next meal was coming from.

Reaganism and Thatcherism were in full swing on both sides of the Atlantic. Looking back it is not surprising that in 1988 a young Oliver Letwin, an acolyte of UK Prime Minister Margaret Thatcher, and today a senior Conservative MP, was advising the Saskatchewan Government on privatizing the province's Crown Corporations. Roy Romanow, the then Leader of the NDP Opposition (later Premier) referred to him in the Legislature 'as the real Premier of Saskatchewan ... and he is from the United Kingdom and he wants to destroy Saskatchewan by privatization' (Hansard, 1988).

Any direct influence Letwin may have had on welfare policy is not known though the then Minister of Social Services was denying 'that poverty was a problem in Saskatchewan even as food banks flourished and children went to school hungry' (Mackinnon, 2003). Romanow went on to say that 'the ideology (of privatisation) has been proven to be old, to be out of touch. It never worked: it led to the Great Depression of the great 1930s, the Great Depression which caused so much misery' (Hansard, 1988).

Indifference today

Privatization of social welfare always goes hand in hand with denial and indifference. As Suzi Leather pointedly noted in her 1996 Caroline Walker Lecture on the rise of modern malnutrition and food poverty in the UK 'the link between poverty and ill health was persistently denied under Margaret Thatchers's leadership' (Leather, 1996), and has continued onwards. Similar indifference is found today in Australia where Sue Booth has noted surveys seeking to measure food insecurity face challenges as there is little political appetite for acknowledging that a problem exists in the first place (SB: email communication 28.12.16).

Echoing comments made three decades ago in Canada, UK Conservative politicians in 2013 were denying hunger and the need for food charity. Lord Freud, a DWP minister, claimed that 'more people were going to food banks because the food was free, thereby triggering "almost infinite demand"' (Butler, 2014). It was a supply not a demand issue. Meanwhile Chris Steward, a senior Conservative Councillor in York asserted there was 'no real poverty in Britain and people should not donate to food banks ... they were an insult to starving people around the world, and that donating to them allowed recipients to spend more money on alcohol and cigarettes' (The Press, 2013).

Dimensions of food poverty

It has been recognized that 'hunger is a difficult term to conceptualize but three dimensions are typically distinguished: biological, social and economic' (Riches and Silvasti, 2014). These are reflected in Box 2.1 which presents overlapping concepts and terms used to define food poverty, food insecurity and hunger drawn from the literature and international population surveys.

BOX 2.1 DIMENSIONS OF FOOD POVERTY: FOOD INSECURITY AND HUNGER

Food poverty

- 'the inability to acquire or consume an adequate quality and quantity of food in socially acceptable ways, or the uncertainty that one will be able to do so' (Dowler, 2003);
- an issue of access to healthy food and its affordability (Silvasti and Riches, 2014);
- a useful synonym for food insecurity (Dowler, 2003).

Food insecurity

- 'the inadequate or insecure access to food due to financial constraints' (Tarasuk, 2016);
- 'the limited or uncertain availability of nutritionally adequate and safe foods or limited or uncertain ability to acquire safe, nutritious food in socially acceptable ways (e.g., without resorting to emergency food supplies, scavenging, stealing or other coping strategies)' (USDA ERS, 2016; see Taylor and Loopstra, 2016);
- 'inability to afford to eat meat, fish or protein every other day' (Eurostat, 2016a)
- 'limited access to food, at the level of individuals or households, due to lack of money or other resources' ... '*mild*: worrying about ability to obtain food; *moderate*: compromising quality and variety of food; and reducing quantities/skipping meals; **severe**: gone entire days without eating ... experiencing hunger' (FAO, IFAD, UNICEF, WFP and WHO, 2017).

Hunger

- the uneasy and painful personal sensation caused by lack of food (Anderson 1990);

- *undernourishment*: individuals consuming less than their calorific/protein/ nutrient requirements for an active and healthy life (FAO, IFAD &WFP, 2015);
- a condition in which people do not get enough food to provide for the nutrients for fully productive, active and healthy lives (BW, 2009);
- 'an individual level physiological condition that may result from food insecurity' caused by a 'household-level economic and social condition of limited or uncertain access to adequate food' (USDA ERS, 2016a);
- '*very low food security*' (USDA ERS, 2016a) – 'federal euphemism for *hunger*' (Poppendieck, 2014a);
- '*severe food insecurity*': '… at the most extreme go day(s) without food' (Tarasuk *et al*, 2016);
- '*severe food insecurity*' – 'experiencing hunger' FAO-VOH (2016).

Food access and food security

- 'secure access at all times to sufficient food' (Maxwell and Frankenberger, 1992, p.8);
- 'when all people, at all times, have physical, social and economic access to sufficient, safe and nutritious food that meets their dietary needs and food preferences for an active and healthy life' (FAO, 2009);
- affordability: sufficient money to shop for food in socially acceptable ways.

Most important are the purposefully drawn links between the issue of food poverty – *food insecurity* and *hunger* and that of achieving food security – through *food access* and *affordability*. In so doing hunger is understood not only as an economic but also a moral and political question. In other words whether people receiving food aid really are hungry reflects an ideological standpoint dependent on one's moral values. It underlines the point that hunger is a thoroughly political question (see Riches and Silvasti, 2014).

Rich world hunger is slowly attracting the attention of civil society though indifferent governments still lag far behind. Interestingly in the USA during the rediscovering of poverty in the 1960s, hunger became the rallying cry for action. Janet Poppendieck, emerita professor and prominent food policy expert has commented 'anti-poverty activists made a strategic decision at that time to pursue reform and expansion of food programmes, rather than the more adequate cash assistance (as in European style welfare states) that might have made such programmes unnecessary'. She notes, quoting journalist Nick Kotz (1984, p.22) that advocates for the poor identified hunger 'as the one problem to which the public might respond. They reasoned that "hunger" made a higher moral claim than any of the other problems of poverty. Federally supported housing and jobs programs could wait; but no one should go hungry in affluent America' (Poppendieck, 1997, p.135). This proved correct as evidenced by the vast expansion and array of public and charitable feeding programmes which comprise the USA's modern food and nutrition safety net (see Poppendieck, 2014a, pp.176–178; 181), and the origins of today's food bank nations.

Food is the elastic item in the household budget given that rent and fuel bills have to be paid. Hunger is not far behind, a stark reality and lived experience of millions of people in the rich world. Yet, it struggles for public attention particularly when housing costs and homelessness, rightly so, are equally pressing. Hunger needs to be named for what it is. While developing evidenced-based policy informed by the concepts of food poverty and food insecurity is essential, people who lack the money to put food on the table do not call themselves food insecure or food poor, they experience and speak of themselves as being hungry. For that reason this study refers to domestic hunger in the rich world.

However that is not to forget what Jessica Powers, the former director of Why Hunger's Nourish Network for the Right to Food in the USA has commented: 'nearly five decades of food banks have failed to solve the problem of hunger because it frames the problem as a lack of food (hunger), rather than a lack of income (poverty) and the solution as distribution, not structural change' (Powers, 2015a), a critical observation addressed later in the book.

Prevalence of food insecurity

The global struggle to eradicate hunger and achieve food security has been shaped by a series of World Food Summits (1974 and 1996, 2002 and 2009) and UN FAO policies and actions. The 2000 Millenium Development Goals (MDGs) achieved the halving of extreme poverty and hunger by 2015, the proportion of people living on less than $1 a day, but 795 million people, one in eight worldwide still remain hungry (UN Fact Sheet, 2015). Understandably, the prevalence of food insecurity in the rich world was barely on the agenda.

In any event the count of those going hungry in the OECD is difficult to determine. There is as yet no international and commonly agreed standard measure of household food insecurity, although one is now in the making. The national and international surveys in use offer varying estimates but lack comparability, owing to differing definitions, methodologies and political indifference. Nevertheless when undertaken they direct attention to the prevalence of food poverty and hunger within and across OECD states, including associated risks and causes. They all shed differing degrees of light on the issues of food access, affordability and income poverty.

With this in mind it is essential first to consider the national monitoring of the prevalence of household food insecurity within OECD countries, and the lack of it, as it provides a context for understanding why agreed standard measures are necessary both nationally and internationally.

National indicators and prevalence

At a seminar in 2014 at the Centre for Food Policy at City University in London, Patrick Butler, *The Guardian*'s Social Policy Editor, asked me what would be the first thing I would do about food poverty in the UK if wearing then Prime Minister David Cameron's shoes. In light of his colleague Owen Jones having written that 'hundreds of thousands can no longer afford to feed themselves' in Britain (Jones, 2014), I suggested the PM's first priority should be to make a national official count of food insecure and hungry people. Jones's figures, while indeed shocking, were likely based on food bank usage and as such would be significant underestimates.

My response was a little smug. Even though Canada and USA have comprehensive official national household food insecurity data, even their governments continue to ignore it. Country specific national household measurements are essential for providing valid and reliable data necessary for determining the prevalence of food insecurity, its degrees of severity and who is at risk so as to inform evidence-based public policy for addressing issues of food access and affordability. However within the OECD such robust population-based surveys are a rarity. Food insecurity data which accounts both for the duration and the severity of people's inability to access food is not routinely collected. Instead governments may consult inferential national poverty data or household food expenditure studies. What then may be learned from the North American surveys?

Food insecurity in North America

National population based household surveys of food insecurity have been routinely conducted in the USA since 1995, in Canada from 2005, and Mexico since 2008.

TABLE 2.1 Food insecurity, headcount by degree of severity. National Surveys,[i] USA and Canada 2012[ii]

	Prevalence		Degree of severity
	Millions	%/Pop	
USA/USDA	48.9m	(15.9%)	food insecure
	(31.7m)	(10.3%)	– *low food security*
	(17.2m)	(5.6%)	– *very low food security (hunger)*
Canada/CCHS	4m	(11.6%)	food insecure[iii]
	Headcounts		– *marginal food insecurity*
	Unavailable		– *moderate food insecurity*
			– *severe food insecurity (hunger)*

Source: *Household Food Security in the United States in 2015* (USDA ERS 2016b); *Household Food Insecurity in Canada 2012* (PROOF, 2014)

i   Individuals (adults and children under 18 yrs) in food insecure households, USDA & CCHS
ii Last year for comparable Canada/US data
iii See discussion below re degrees of severity.

USA

The United States Department of Agriculture's (USDA) survey is of 45,000 households evaluating levels of food insecurity experienced over the previous 12 months. A food secure household is evaluated as either *high* – 'all members at all times have access to enough food for an active, healthy life'; or *marginal* – if at a minimum nutritious and safe foods are readily available; and there is an assured ability to acquire food in socially acceptable ways. Importantly this means 'without resorting to emergency food supplies, scavenging, stealing or other coping strategies' (USDA ERS, 2016b).

By contrast food insecurity, as Table 2.1 indicates, is the limited or uncertain availability of safe and nutritious foods and the limited or uncertain ability to acquire it in socially acceptable ways. Food insecure households are reported as experiencing either *low* or *very low food insecurity* (see USDA ERS, 2016b). Accessing a food bank would be a marker of a food insecure household and *very low food security* an indicator of hunger (Poppendieck, 2014a).

Table 2.1 also indicates the prevalence of food insecurity in 2012 in the USA. 48.9 million Americans (15.9% of population) were reported as food insecure in 2012, of whom 31.7 million were struggling with 'low food security' (8.7%) and 17.2 million (5.6%) suffered from 'very low food insecurity' (USDA ERS, 2016c). Certainly as a result of an improving economy the prevalence of food insecurity in the USA has declined to 42.2 million (13.4%) yet it still remains above the level prior to the onset of the Great Recession in 2008 (USDA ERS, 2016b). Hunger remains deeply embedded in the world's strongest economy, ironically a country with an unparalleled public and charitable food safety net.

Canada

Table 2.1 shows some of the difficulty in comparing numbers, rates and different experiences of food insecurity between the two countries. As Valerie Tarasuk has advised one issue is the slightly different coding systems for the two surveys. In Canada the monitoring and national measurement of food insecurity is conducted as a module within the Canadian Community Health Survey (CCHS). It is a population-based health survey of 18,000 households of which the sample design is jointly developed by federal, provincial and territorial representatives. Whilst the CCHS food insecurity measures have been adapted from those in the States its auspices are those of public health rather than agriculture as in the USA.

Other differences are that each of the 18 questions in the adult (18 yrs+) and child survey references the lack of money as the primary reason for: worrying about running out of food *(marginal food insecurity* – not so in the US where you would be considered food secure); relying on low-cost food and unable to afford balanced meals *(moderate food insecurity)*; and cutting the size of meals, skipping meals, not having enough to eat and at the extreme going days without eating *(severe food insecurity)*, in other words going hungry (see CCHS, 2016; Tarasuk et al,

2014). In short, in Canada household food insecurity is defined as 'inadequate or insecure access to food because of financial constraints' (Tarasuk et al, 2014), focusing on food access and affordability.

In 2012, in Canada, 4 million people (11.6% of the population) of whom 1.15 million children lived in food insecure households (see Table 2.1), an increase of 606,500 individuals since 2007 (Tarasuk et al, 2014). Whilst food insecurity is most extreme for Aboriginal peoples living in Northern Canada's third world conditions, even the CHHS survey is an underestimate. It excludes individuals living on First Nations reserves or Crown Lands, full-time members of the Canadian Forces, persons in prisons or care facilities, and the homeless. Another is that in Canada, provinces and territories may and do opt out of participating in a particular survey year.

Despite the fact that 2012 was the last year when a complete Canadian national survey was undertaken as Tarasuk has written 'although there has been rigorous measurement and monitoring of household food insecurity in Canada since 2005, the problem has not abated. In fact, it has grown and persisted in every province and territory' (ibid). In fact it is a public health crisis. The least one can say is that in the North America valid and reliable estimates of food poverty are routinely collected.

Mexico

It is also important to recognize that Mexico as an Upper Middle Income OECD member state monitors food insecurity on a regular national basis through two national surveys. The population based National Survey for Health and Nutrition (ENSANUT) measures household food insecurity and has done so on a six-yearly basis since 1988. Meanwhile the National Council for the Evaluation of Social Policy (CONEVAL) using the Food Security Mexican Scale (EMSA) has monitored individual food insecurity on a two-year basis since 2008. Both surveys categorize food insecurity as low, or moderate and severe with severe meaning lack of access to food. In 2014, 28 million Mexicans, 23.4% of the total population lacked access to food (see Castrejon-Violante, 2017a; CONEVAL, 2015).

Missing data in OECD member states

The irony in both the USA and Canada is despite the robust and systematic collection of national food insecurity statistics, a prerequisite for evidence-based and effective public health and social policy, their indifferent governments ignore their own data. Domestic hunger continues. Inadequate minimum wage and income assistance policies are unaddressed. However, the key point is that given the lack of national monitoring of food insecurity in other OECD countries, the USDA, CCHS and Mexican surveys provide official standard measures of food insecurity for possible adaptation elsewhere.

Lacking such data EU member states either infer the extent and depth of food poverty from national poverty or household expenditure statistics or rely on the

EU-Survey of Income and Living Conditions measure of severe material deprivation whether a person 'has eaten a meal with meat, chicken or fish or other protein
rich nutrition every second day'. While implemented annually in the UK since
2005 (see FRC, 2016, p.151), this lonely indicator hardly seems a sufficiently
robust measure of food poverty as it fails to explore the severity of food insecurity
and hunger.

In the UK where food poverty is the preferred term (APPG, 2014; Dowler,
2014), people's inability to feed themselves has been inferred from the fact that 13
million people, 21% of the population, live below the poverty line with five million
in workless households (Dowler, 2014). The systematic national measurement of
food insecurity has not been a government priority.

A similar scarcity of robust food insecurity data exists in Europe where in 2014,
45 million people are possibly unable 'to eat fish, meat or other protein rich
nutrition every other day' (Eurostat, 2016a, p.151), 'according to Caritas Europe,
79 million people are food insecure' (Escajedo San-Epifino and De Renobales
Scheifler, 2015, p.18). Whilst the EU-SILC (2014) reports 45 million as being food
insecure, Caritas reports 79 million people, 15% of the EU population live below
the poverty line (less than 60% of average earnings), of whom 16 million receive
food aid (Caritas Report, 2014). Who is right? The problem is the lack of comprehensive and routinely collected national food insecurity data by each EU
member state which captures the degree of severity and temporal nature of being
without food. Even Finland, a proud Nordic welfare state, lacks timely official data
of those receiving food aid (Silvasti and Karjalainen, 2014). Similarly, in Australia as
Sue Booth has noted 'the lack of a comprehensive food insecurity measure and
regular monitoring means it is difficult to acknowledge the size of the problem'
(Booth, 2014, p.17). The situation is much the same in New Zealand where as
Michael O'Brien agreed the NZ government does not attach much priority to
food insecurity where the response is mostly a charity model (MO'B – email
communication 10.7.17).

The lack of comprehensive and robust national food insecurity data to inform
public policy in the majority of OECD member states is particularly troublesome.
Even those states such as Canada and the USA which do have reliable data are
seemingly content to leave the hunger problem to the charity economy.

International indicators and prevalence

What then can be learned from the international monitoring of food insecurity
about domestic hunger in OECD member states?

FAO: State of Food Insecurity in the World

In 2014 the FAO's State of Food Insecurity in the World report found 15 million
people living in developed regions to be experiencing 'undernourishment' (FAO-
FISW, 2015). These were individuals consuming less than their calorie requirement

for an active and healthy life. In the context of global hunger 15 million is only a tiny proportion (2%) of the 795 million undernourished people worldwide (10.9% of the global population), and likely a measure easily passed over by policy analysts in advanced industrial economies. Yet, it hints at an issue of food poverty and domestic hunger in the rich world. Certainly this is suggested by the North American national surveys, but also by EU and more recent FAO measures.

EU Survey of Income and Living Conditions (Eurostat, 2016a)

The EU–SILC study uses a particular measure of food insecurity within the framework of poverty and social exclusion: monetary poverty after social transfers; very low work intensity; and severe material deprivation. It asks a single question about household food insecurity: whether people are 'unable to afford to eat meat, fish or protein every other day'. People suffering from severe material deprivation are those 'living in conditions greatly constrained by a lack of resources and cannot afford at least four of the following: to pay their rent or utility bills or hire purchase instalments or other loan payments; to keep the home warm; to pay unexpected expenses; to eat meat, fish or other protein-rich nutrition every second day; a week long holiday away from home; to own a car, a washing machine, a colour tv or a telephone' (Eurostat, 2016a, p.151).

In 2014 nearly 122 million people (23.7%) in the EU were at risk of poverty and social exclusion including 45 million people (9%) who 'were living in conditions severely constrained by a lack of resources' (Eurostat, 2016a, p.158), in other words going hungry. Despite a modest decline in poverty and social exclusion since 2012 the provisional 2015 rate still remained above that of the 2009 low point (23.3%), nearly one in four of the EU population.

However, a single EU food insecurity indicator is hardly a robust measure of the prevalence of food insecurity. The survey also excludes specific population groups, those living in collective households, institutions and remote regions (Eurostat, 2016b, p.5). Nevertheless attention is drawn to income poverty and the problem of affordability and food access.

FAO Food Insecurity Experience Scale (FAO-VOH, 2016)

The *Food Insecurity Experience Scale* (FIES), being developed by the Voices of Hunger project under the aegis of the FAO, is a step towards developing a standard measure for estimating comparable prevalence rates of food insecurity experienced by individuals or households in more than 140 countries (rich and poor). The survey is focused on adults' direct experiences of food insecurity.

As indicated in Table 2.2 the preliminary findings for 2014 show 145 million individuals or 11.1% of the total OECD population experiencing moderate and severe food insecurity (FAO-VOH, 2016). This is an estimate of the percentage of people who have been forced to reduce the quality and variety of food they eat and might have been forced (due to lack of money or other resources) to also start reducing its quantity.

TABLE 2.2 Prevalence of moderate and severe food insecurity in the OECD as a percentage of total population, provisional estimates, FAO-VOH, 2016[i]

OECD states (millions)	Individuals (millions)[ii]	Percentage of population[iii]
High Income		
Europe (25)	41.0	8.6%
(UK)	(8.4)	12.9%
Australia	2.8	11.7%
Canada	3.0	8.3%
Chile	2.6	14.5%
Israel	0.4	5.0%
Japan	3.7	2.9%
Korea	3.3	6.4%
New Zealand	0.5	11.1%
USA	33.2	10.2%
	90.5m	
Upper Middle Income		
Mexico	36.7	26.9%
Turkey	17.4[iv]	n/a
Total OECD	**145m**	**11.1%**

i   Derived from Appendix, Table A-1, pp.36–39, FAO-VOH, 2016.

ii Estimate of number of individuals in the total population living in households where at least one individual aged 15 or more is classified as food insecure. NB data on which these estimates are based are not from national sources.

iii Europe 475m; OECD 1.3b. see FAO Food Balance Sheets, Population_2015_FAOSTAT-3.csv.

iv Turkey estimate based on Adults 15yrs and older.

As an experienced-based module the eight question metric gauges 'the severity of the food insecurity condition of individuals and households' (FAO-VOH, 2016, fig. 2.1, p.3), food insecurity is categorised as

mild: worrying about ability to obtain food;
moderate: compromising quality and variety of food, reducing quantities and skipping meals; and *severe:* experiencing hunger (ibid).

It must be noted that the preliminary findings from the FIES-World Gallop telephone surveys are based on small sample sizes and are not directly comparable to the those of the US and Canadian food insecurity surveys. As Valerie Tarasuk has commented they compare unfavourably with the national Canadian survey with 'a very carefully constructed population based sample of 65,000 per year' and 'a food insecurity measure based on 18 items' (VT – email communication, 16.5.16).

However there are two key points. Firstly, the FIES surveys do point to wide-spread food insecurity in wealthy food bank nations. Secondly, they bring to global and national attention the importance of developing a comprehensive, valid and reliable standard measure of food insecurity. The VOH-FIES project is playing an invaluable role in putting rich world hunger and food insecurity on the map.

Food bank usage

In the absence of standard and comprehensive monitoring of food insecurity in many countries national food bank usage has become the proxy measure. The numbers of adults and children resorting to food banks are now accepted as the public face of domestic hunger. Yet glossy on-line websites and annual reports reporting the millions of meals provided, tons of surplus food redistributed, donations received and millions of people fed are deceptive.

With access to national food bank and EU food aid data (2010–2012), Ugo Gentilini of the UN World Food Programme estimates that up to 60 million people were food bank beneficiaries in 24 high-income countries comprising Canada, the USA and 22 EU states. Yet the 'actual number of beneficiaries' he states could be 'considerably higher' (Gentilini, 2013). For example Germany, the Netherlands, Sweden and the UK are not included in the EU count; the US data only represents 80% of food bank outreach; and in some countries data does not exist (ibid). Finding robust, comparable and consistent time food bank data is challenging.

Caution is therefore needed in its interpretation and also recognizing that the number of people turning to food banks is a poor indicator of the prevalence of food insecurity. As Table 2.3 shows food bank usage in the OECD significantly underestimates the prevalence of food insecurity and domestic hunger.

The fact that 46.5 million Americans turned to a food bank in 2014 when 48.1 million were judged food insecure by the USDA survey may let us believe that Feeding America, the peak US charitable food bank organization, is single-handedly managing the problem, while feeding public perceptions in other OECD countries that food charity is resolving the hunger issue.

TABLE 2.3 Food insecure individuals and food bank recipients. North America, EU and OECD, survey estimates 2012/2014

Country/Region	Food insecure individual (million)s	Food bank[i] recipients	Associations
USA (2014)	48m	46.5m	Feeding America
Canada (2012)	4m	860,000	Food Banks Canada
EU (2014)	45m	5.7m	FEBA
OECD (2014)[ii]	145m	6.8m	GFN (OECD plus)

Source: USDA (2016b); FA (2014); PROOF (2012); FBC (2012a); FEBA (2016); Eurostat (2016a); FAO (2016); FAO-FIES (2014); GFN (2016)

i   Nationally collected food bank data.

ii FAO-FIES scale.

Yet, as Janet Poppendieck has noted 'if you type "fighting hunger in America" into almost any search engine' it will take you to food charity websites and corporate philanthropies while bypassing government programmes 'which provide the vast majority of the nation's food assistance' (Poppendieck, 2014a). Interestingly it has been estimated that 40% of Feeding America's recipients have eligibility for federal food assistance but have not applied for SNAP benefits (Gentilini, 2013). Leading US anti-hunger organization Bread for the World has also estimated that only 'one in 20 bags of food assistance came from a charitable organization with Federal nutrition programs providing the rest'. Food charity spent $5.2 billion feeding the poor compared to $102.5 billion by the Federal state (BW, 2014).

In Canada the comparative statistics are for 2012, the last year when complete national food insecurity data is available. There is clearly a significant gap between the 4 million individuals found to be food insecure by the Canadian Community Health Survey and the 882,118 food bank users per month reported in Food Banks Canada's HungerCount Survey for that year (PROOF, 2014; FBC, 2012a).

In other words food banks were assisting just a little more than 1:4 of hungry Canadians. Importantly, while 'social assistance recipients comprised 52% of food banks users' in 2012 (FBC, 2012a) they counted for only about one quarter of the food insecure population. The working poor, in contrast, comprised the majority of food insecure Canadians but only 12.5% of FBC's clientele' (Riches and Tarasuk, 2014). Moreover research with people who use food banks or eat in charitable meal programmes indicate that many still go hungry (Hamelin et al, 2002; Loopstra and Tarasuk, 2012). Certainly, food banks provide emergency food relief and FBC routinely acknowledges that it cannot solve food insecurity but food bank usage significantly underestimates the prevalence of food insecurity (Loopstra and Tarasuk, 2015).

The European and Global Foodbanking associations' data likewise indicate the huge gaps between the numbers officially recognised as hungry and those turning to foodbanks. Still, the charitable food bank counts, even in the USA and Canada with robust national food insecurity data, allow indifferent governments to rest easy falsely believing the hunger problem is being addressed and shielding their eyes from the persistence of domestic hunger.

Persistence of domestic hunger

'Undernourishment', 'inability to afford a nutritious meal every other day' and the experience of 'moderate' and/or 'very low/severe food insecurity' reflect differing definitions and measures of food insecurity and hunger and their prevalence in the rich world, including the problem of constructing reliable and consistent indicators. This says nothing about the difficulties of interpreting the findings from surveys, national and international, based on different years using a variety of research methodologies.

Despite these inconsistencies and shortcomings, the sheer numbers of people unable to put food on the table reveal that domestic hunger in the rich OECD is pervasive and entrenched in our abundant but deeply unequal societies. Despite the variation between OECD countries regarding the numbers and prevalence of those unable to feed themselves, one could well argue that US and Canadian food bank usage data alone is evidence enough of the profound nature of the crisis and human suffering. However as is shown in North America and elsewhere in the OECD (Table 2.3) food bank figures are merely the tip of the iceberg

In Europe the wide discrepancy between the 45 million people officially declared in 2014 as unable to afford a quality meal every second day (in other words food insecure and hungry) and the 5.7 million fed in 2015 by the FEBA's charitable food banks in 23 EU states is glaringly obvious (FEBA, 2016b). Food banks were reaching little more than 12.6% of Europe's hungry poor. The UK tells a similar story. As the Food Research Collaboration states ' where 13 million people are considered to be at risk of poverty and last year more than a million received food parcels from Trussell Trust food banks, the number of adults and children who are food insecure remains a mystery', even noting the work of other food bank associations (FRC, 2016).

In the OECD the gap between the prevalence of food insecurity and food bank usage is starkly illustrated by the provisional estimates of the FAO-FIES survey conducted in 2014 (see Table 2.2). Whilst the FIES survey instrument is not yet fully developed, preliminary estimates show 145 million individuals to be experiencing moderate or severe food insecurity (again likely underestimates) in 34 OECD member states. Meanwhile in the same year the Global Foodbanking Network was feeding 6.8 million people in 32 countries around the world of which eight are OECD member states: Australia, Canada, Chile, Israel, Mexico, Turkey, the UK and the USA with Feeding America is a partner (GFN, 2016a).

While the national and international measurements of food insecurity have their limitations such data collection is imperative as basis for evidence-based policy making. Their varying estimates depict a landscape of shameful domestic hunger in the rich world in a context of widespread income poverty and growing inequality. They illuminate the lack of affordability and daily access to food in normal and customary ways for millions of people posing questions about food, public health and social rights in supposedly 'food secure' nation states.

Food insecurity data is essential for understanding that for too long food charity has been nibbling away at the deeply rooted systemic problems of food poverty (and food waste) while the indifferent State muddles along uninterested in producing valid and reliable estimates of the hungry poor, nor even bothering to ask who is hungry and why.

Who is hungry and why

As the differing measures of food insecurity and domestic hunger in the OECD tell us income poverty, material deprivation and widening inequality are the primary causes of domestic hunger. People, working or not, lack sufficient cash to pay the rent, heat their homes *and* feed themselves and their families, even assuming they have no other costs of living and are not homeless on the street. People simply lack the money to make ends meet.

Working poor and unemployed

Hungry people are the income poor, the under and unpaid members of the 'precariate'. They are people whose inadequate wages condemn them 'to earning their poverty' (Jones, 2017) who are one pay cheque away from putting food on the table. In the UK many are women (Dowler, 2014, p.161). In Canada research shows the majority of the food insecure are the working poor (Tarasuk et al, 2014). Politically accepted as 'the deserving poor', they are the exploited underemployed with strong work ethics seeking to discharge their family obligations and social responsibilities by clinging tenaciously to the thinly stretched lifeline of sub-poverty incomes and precarious employment.

They are also the unemployed and homeless, the undeserving poor claiming benefits who have been made surplus to the requirements of the labour market. Their eligibility for unemployment insurance, social security benefits, food stamps, housing support and disability payments has been questioned, delayed or not granted. Financial assistance or food assistance when received is often so low, that food is unaffordable. Even school meal programmes, if available, may not keep child hunger from the door.

The risk of hunger primarily depends on one's perch on the economic ladder: good jobs/bad jobs; full-time/part-time; zero-contract hours; fruitlessly looking for work; jobless; inadequate employment income or social security benefits; benefit delays; rent arrears and unmanageable debt; or finding oneself homeless. The food insecure are people whose 'vulnerability is tightly entwined with labour conditions' (Riches and Tarasuk, 2014). Whether in or out of work hungry people are all up against an unjust wage and social security benefit system.

Vulnerable populations

The risk of hunger in the OECD world is determined by a complex set of interacting demographic, social and cultural factors: gender; age (young and old); family composition (size, single parenthood and single hood); race and ethnicity; Aboriginal and Indigenous status; immigrant and refugee status; disabilities; education; physical and mental health; addictions and substance abuse; a home or the lack of it and the remoteness of where one lives. Finding oneself vulnerable in any of these interconnecting spaces and left unprotected by social legislation and a complacent State makes feeding oneself and one's family a daily nightmare.

The voices of hungry people and their many personal narratives and testimonies are found not only in the academic literature but in a range of reports and documentaries by leading human rights, food policy and social justice advocacy organizations. They reveal not only the human experience of hunger in today's wealthy societies but the intransigence of the market economy when unregulated excessive profit-making takes precedence over the neo-liberal State's obligation to ensure the right of everyone and their families 'to an adequate standard of living including adequate food, clothing and housing, and to the continuous improvement of living conditions' (*ICESCR*, Article 11, 1966), and their ability to meet their family and social obligations and to participate in and contribute to society.

Income poverty, inequality and social injustice

If these are the economic and social risk factors closely associated with rich world hunger one does not have far to look for underlying structural causes. Ironically, in those wealthy OECD nations where food is supposedly cheap, its affordability depends on adequate wages, income security and inclusive health and social policy. Food insecurity is clearly rooted in unfair income distribution and widening inequality. Whether food prices rise or fall, the power and wealth of the 1% are constant reminders of the gap between the rich and poor. Since the early 1980s repeated economic recessions (most recently the Great Recession of 2007–2009), significant increases of low-waged precarious part-time employment and joblessness has left millions dependent on charitable food hand outs.

This is morally unacceptable in a world in which Oxfam reports the globe's richest eight people have the same wealth as the poorest half of humanity (Oxfam, 2017) Even in tolerant and compassionate Canada by midday on January 3rd 2017, the Canadian Centre for Policy Alternatives reported that the 100 top CEOs earned what the average worker would have to work full-time all year to make (Mackenzie, 2017). Astoundingly, five Canadian families are as rich as 30% of the population (CB, 2017). Nor did this study take into account those whose sub-poverty social assistance benefits left them unable to pay the rent, heat the home let alone feed themselves. All the more reason for governments to check the facts about who is hungry and why.

The legacy of less eligibility

In the past three decades the 19th-century principle of less eligibility which historically informed the Victorian poor laws (de Schweinitz, 1961; Guest, 1997) has found new life informing the neo-liberal administration of punitive welfare policies in today's food bank nations. Less eligibility stipulates that no one on relief should receive more than the lowest paid wage earner, in other words subsistence level (or less) welfare benefits. It is about deterrence, disciplining labour and the poor by ensuring downward pressure on wage levels for the under and precariously employed. In social policy its function is to direct the unemployed to accepting low wage and zero contract hours employment. It is the guiding principle of neoliberal welfare reform.

As Karl de Shweinitz reminds us the 1834 Victorian Poor Law reforms 'placed the burden of destitution upon the shoulders of the individual. Poverty was regarded as essentially an indication of moral fault in the person requiring relief. He was held little short of exclusively responsible for his condition' (de Schweinitz, 1961, p.126). What today we would call 'blaming the victim'. As for the role of the state he wrote 'the idea of social obligation was not conceived in the thought of the times' (ibid). Fast forward nearly two centuries, and what have we learned? Millions of people in the excessively affluent OECD are begging for food while governments look the other way.

Subsidiarity

Neoliberal policy-making also embraces the organizational principle of subsidiarity. It highly values decentralization and the making of political decisions as close to the local level as possible, for example the 19th-century practice of the parish administration of poor law relief. This fits well with today's austerity driven governments committed to fiscal constraint and always flying the flag of lower taxes. Downloading financial costs and social welfare responsibilities to lower levels of government (e.g., states, provinces, municipalities) and then outsourcing society's bad risks to under-resourced and unaccountable charitable food banking shows how much affluent nation states have turned back the clock of social reform.

While constitutional divisions of power for health and social policy vary between unitary and federal OECD states nevertheless the expectation that feeding hungry citizens is the responsibility of local communities is to be found in the majority of food bank nations. Food charity is a convenient fit for neo-liberal governments with no political will for addressing the complex issues of widespread food poverty within their jurisdictions.

Moral vacuum of neo-liberalism

The moral vacuum at the heart of neo-liberalism is the misleading and divisive ideological message preached by governments everywhere that economic growth will bring shared prosperity to all. Instead we have widespread domestic hunger. Politicians constantly repeat its central mantra that work is the best and only social policy and way out of poverty (Silvasti and Riches 2014). If only this were true, but precarious employment, low waged work and unemployment to say nothing of robust public health and social policies do not advance social mobility nor ensure access to food for vulnerable populations.

The political elite's 35-year embrace of financial deregulation, privatization, smaller government has restructured and enfeebled welfare states. The austerity mindset of financial restraint, social spending cutbacks, welfare reforms eroding entitlements and income-based social safety nets coupled with the maxim of lower taxes and regressive taxation could only have one result: more severe levels of poverty, material deprivation and domestic hunger. The State's corporate embrace of the market has broken the social contract (see Jones, 2014).

Undoubtedly within the privileged high and upper-middle income OECD member countries there are significant differences between the social values and the financial and resource capacities which underpin their respective welfare states, and specifically for ensuring access to food. Yet it is salutary to think that among the social democratic and universal welfare states of the Nordic countries, the strong conservative and corporatist as well as less developed welfare regimes found across Europe, the 'anglo-saxon' liberal and residual welfare states of Australia, Canada, New Zealand, the UK and the USA (see Alcock and Craig, 2009) and those of Chile, Mexico and Turkey, food banks are now part of the charity economy.

Reflections

Bearing in mind the differing national and international meanings and measurements (and the missing data) of food poverty, food insecurity and hunger it is abundantly clear that millions of people (140 million is likely a considerable underestimate) in the rich OECD world cannot afford to put food on the table. Domestic hunger is pervasive and entrenched. It is not only the jobless and those dependent upon social security or food stamps who go hungry, many are the working poor. Women, children, ethnic minorities, Indigenous peoples, asylum seekers and refugees are among those severely impacted.

With the exception of North America OECD member states lack standard and robust national measures of food insecurity. Their absence signifies the State's shameful indifference to public begging for food. Even the USA, Canada and Mexico with official and reliable data outsource domestic hunger to charity.

As Tiina Silvasti, professor of social and public policy, and I have previously written 'income poverty, par excellence, is the primary cause of domestic hunger and increases the demand for food aid' (2014), aided and abetted by unfair income redistribution, widening inequality and social injustice driven by 35 years of neo-liberal ideology and punitive welfare reform. Economic growth and work alone is not the short cut nor even a long-term escape out of poverty. Wherever you live, whatever the state of the market economy and the welfare state, you need sufficient money in your pocket, from wages or income assistance, if you wish to eat, be healthy and participate in society.

Instead indifferent governments deny the problem and look the other way leaving the task of feeding the poor to the redistribution of wasted and surplus food. For this reason when considering the rise of food bank nations it is crucial to understand rich world hunger as a political matter central to human rights and public policy.

3
RISE OF FOOD BANK NATIONS

Historically there is a rich heritage of charitable giving dating back to Victorian times and long before which continues to this day. In all OECD high income countries charities such as the Salvation Army, Caritas, St Vincent de Paul, the Red Cross, missions and settlement houses, housing and shelter organizations are reminders of this. When income assistance and public social services are missing in action, faith-based charities of different religions and denominations and community welfare organizations working at the street level have always stepped in.

Whatever the causes and as best they can emergency food relief organizations such as soup kitchens and meal programmes have long been at the forefront of efforts by concerned citizens, religious and community non-profit organizations to provide for those who have slipped through threadbare social safety nets. Today's charitable food banks, markers or beacons some might say of newly emerging food safety nets in the world's rich nations, did not spring out of nowhere.

The legacy of charity and food safety nets

In more recent history the breadlines and soup kitchens of the Great Depression are constant reminders of the vital roles played by emergency food charity in the struggle against mass unemployment, poverty and hunger in the 1930s as well as the role of government. In the USA, as Janet Poppendieck reminds us, in the crisis years of the 1930s the Federal Government took the path of the food safety net. Both the Hoover and Roosevelt administrations transferred the food bounty of surplus farm products – food intervention stocks – to feed the jobless poor and their families. It established national school lunch programmes, provided free distribution of food through state and local agencies; and introduced an innovative food stamp programme. In 1933 the Federal Surplus Relief Corporation (FSRC) was created within the United States Department of Agriculture (USDA) to transfer agricultural surpluses to relief organizations to feed the unemployed. It was then seen as a temporary programme and as an 'experimental incarnation of the New Deal' (see Poppendieck, 2014b).

Notably in the post World War II era and against the backdrop of the Universal Declaration of Human Rights (1948) the liberal democracies which had emerged victorious set about the task, some more robustly than others, of constructing their welfare states. These were intended to assure the education, health and social well-being of all their citizens and were grounded in a strong role for government and public policy. A primary goal was to reduce if not eliminate poverty, provide public housing and ensure full employment. Social security such as health and social insurance, family allowances and old age pensions were either strengthened or introduced. As for addressing poverty ('outdoor poor relief') income assistance or welfare benefits provided an essential component of publicly funded social safety nets.

The welfare state in the USA took a somewhat different path where the practice of direct food assistance remained central to public relief policies (see Poppendieck, 1997). Indeed the spirit and intent of the FSRC continues today in the form of the Emergency Food Assistance and Soup Kitchen-Food Bank Program also administered by the USDA which supplies food banks and relief organizations with surplus agricultural commodities – food (EFAP, 2017). In this way the food safety net remains an essential and ever present part of the US welfare state, but as Janet Poppendieck observes 'with a priority on benefits to agricultural producers' (1997), in other words agricultural policy drives social policy.

Strategic decision

Most tellingly she comments that when hunger became a public issue in the 1960s 'anti-poverty activists in the US made a strategic decision to pursue the reform and expansion of food programmes rather than the more adequate cash assistance that might have made them unnecessary' (Poppendieck, 1997). Such an immediate and direct appeal to take on the hunger problem in the land of plenty resonated with the American public. This is a key point. The universal moral imperative to feed hungry people acts as a powerful motivating force. There is little doubt that food banking both within the USA and across the rich world has been driven by practical compassion and feelings of social solidarity with the hungry. In more recent years increasing concerns about the environmental and economic costs of food waste are now also driving the food bank movement.

US food bank origins and institutionalization

Modern day global food banking has it local origins in the United States and the pioneering work of John van Hengel and Robert McCarty who in 1967 established St Mary's Food Bank in Phoenix, Arizona. At that time they were both volunteering at the local St Vincent de Paul Society mission. McCarty was a Catholic deacon and van Hengel a business man, social activist and also a devout Catholic.

From St Mary's Food Bank to Second Harvest

John van Hengel soon became aware that the Society was always in need of a steady supply of food for its soup kitchen. A local woman who was feeding her 10 children from food dumped in grocery store garbage bins suggested to him that 'there should be a place where surplus food could be stored and made available to people who needed it, instead of being thrown out and wasted' (JvH, 2017; McCain, 2003a). This in turn led to his finding volunteers to glean unpicked fruit from backyard trees, to contacting local food stores to donate edible but unsaleable food and to the creation of St Mary's Food Bank, a depot where surplus or wasted food could be collected, sorted, stored and then distributed to social agencies who were feeding the poor.

Warehouse model of food banking

The idea caught on in other US states leading to the founding of Second Harvest and its formal incorporation in 1979 led to the increasing national institutionalization of charitable food banking. This faith-based, grass roots and local response to emergency food needs in Phoenix, Arizona was to spawn the development of the warehouse model of food banking which today spans the United States with 200 hundred 'food storage and distribution centres' and its network of smaller front lines agencies often called food pantries (see FA, 2017a).

Though food banks come in all shapes and sizes in responding to the different needs of urban and rural communities food banking has become a top down approach to feeding hungry people. Food banks perform a number of functions. They secure donations and funding from the food industry and government; the warehouses distribute food and grocery items across its network of social agencies. Food banks distribute many millions of pounds of food a year to hunger-relief charities; they train volunteers and work to improve food security through public education and advocacy (see FA, 2017a).

Food Aid: federal and state support

It should be noted that Second Harvest's incorporation was to receive considerable government support. In 1975 a federal grant helped to establish 18 food banks across America (JvH, 2017) in 1976 it was assisted by the passing of the *Tax Reform Act* which encouraged food companies through tax deductions to donate surplus food to charity (FA, 2017b),

Encouraging those food companies which were still hanging back from making food donations, in 1981 the US Congress passed the *Good Samaritan Act* to protect them from liability and being sued in the event that such foods could prove harmful to the health of recipients (Dey and Humphries, 2015). This was followed in 1983 by the passing of the *Emergency Food Assistance Act* setting up, as noted previously the Emergency Food Assistance and Soup-Kitchen Food Bank Program

(EFAP, 2017) which not only supplied USDA commodities (surplus food) to emergency feeding programmes but authorized grants to states helping with the costs of transporting, storing and delivering food to frontline agencies such as food pantries, soup kitchens and other charities working with the poor (JvH, 2017).

From America's Second Harvest to Feeding America

Working with the federal and state governments America's Second Harvest, today's Feeding America, 'established food banking standards and guidelines' as well as the now publicly accepted practice of 'the acquisition of food from large national manufacturers' and 'businesses were able to cut the costs of disposing unusable but edible food as well as taking tax breaks by helping multiple charities' (JvH, 2017). Growing food industry support, aided further by the passing of the *Bill Emerson Good Samaritan Act* of 1996 (Bloom, 2010) further promoted the national development and institutionalization of charitable food banking.

In 2008 America's Second Harvest was re-branded as Feeding America. In 2017, fifty years after the founding of St Mary's Food Bank, Feeding America is at the centre of a national network of 200 food banks distributing food to 63,000 agencies and is, in its own words 'the nations leading domestic anti-hunger organization, which annually provides food to more than 46 million people' (FA, 2017b). There is no doubt food banking is deeply embedded in American culture.

A 'simple' and 'great' idea

St Mary's Food Bank was a grassroots response to the rediscovery of poverty in the southern United States in the late 1960s. The warehouse model of food banking was seen to be an innovative, effective and coordinated response for recycling surplus and wasted food to charitable agencies feeding the poor. As a centralized method for distributing emergency food relief it was an idea which would quickly spread. On the occasion of John van Hengel's 80th birthday Arizona Senator John McCain in his commendation in the US Senate remarked that the 'food banking idea is simple, but like all truly great ideas, it took the efforts of one man working for a lifetime to reach fruition' (McCain, 2003b).

The public appeal of charitable food banking and support for the food safety net in the US lies in its expression of practical compassion rooted in faith-based and community beliefs – the moral imperative to feed hungry people in the land of plenty. The extensive array of both public and charitable food assistance programmes dating back to the hard times of the 1930s including school meals, food pantries and food stamps has proved solid ground on which to build America's food bank movement. This has been enabled and strengthened by the USDA's public provision of surplus food stocks; the seemingly unwavering support of the heavily subsidized Big Food industry backed by the powerful Agricultural lobby in Congress; Good Samaritan and tax incentive legislation coupled with public grants and private donations; and the Emergency Food Assistance and Soup-Kitchen Food Bank Program.

The ready availability and power of volunteerism and community organizing reinforced the strong public belief that feeding the poor through targeted food assistance and 'hunger' safety nets (Berg, 2008) was more effective than government financial assistance and comprehensive social security. It was an idea which travelled well: charitable food banking was to cross more than one national border.

Crossing OECD national borders

The development of charitable food banking and its role in the US food safety net from the late 1960s onwards was to be instrumental in its national and global spread to wealthy OECD member states in the years ahead. Tables 3.1 and 3.2 presented to aid later discussion, depict the chronology of this expansion between 1967 and 2015.

Table 3.1 focuses on the early linkages and establishment of food banks in North America and Europe from Phoenix (Arizona) to Edmonton (Alberta) and Paris (France), and the development of national and international networks – the European Federation of Food Banks (FEBA) and Global Foodbanking Network (GFN).

TABLE 3.1 50 years of food banking: national origins and global spread of food bank nations, 1967–2017

		First food bank	*National and global spread*	*Founded*
N. America				
USA	1967	St Mary's Food Bank	America's Second Harvest **Feeding America (2008)** – network of 200 food banks	1979
Canada	1981	Edmonton Food Bank	Canadian Association of Food Banks ***Food Banks Canada*** (2006) – 10 provincial associations	
Europe				
France	1984	Paris Food Bank	***European Federation of Food Banks*** – 23 countries (19 EU) and projects	1986
Latin America				
Mexico	1987	Guadalajara Food Bank	***Asociación Mexicano de Bancos de Alimentos*** – BAMX: 56 food banks	1995
Global			***Global Foodbanking Network*** – 32 countries and ***FEBA***	2006
			GFN Food Bank Leadership Institute – global food bank training institute	2007

Sources: Food bank websites and annual reports

1980s – Steps in all directions

North to Canada

North of the border as Canada was struggling with high levels of unemployment created by the deep recession of the early 1980s, food banks were attracting the attention of churches and community agencies. Canada's welfare state was under stress (Lightman, 2005). Skyrocketing joblessness coupled with inadequate federal unemployment insurance and provincial social assistance benefits were leaving thousands of people unable to pay the rent and feed themselves (Hurtig, 1999). As Thatcherite fiscal restraint and social spending cuts bit deeply, frontline charities turned their attention south of the border.

Robert McCarty of St Mary's Food Bank travelled to Western Canada to advise on the setting up of the food banks in Calgary, Edmonton and Regina. Food bank organizers in Montreal and Toronto visited their counterparts in New York and Detroit while some gained inspiration from watching US television depicting the work of food banks and others corresponded with US food banks. The director of the Great Vancouver Food Bank had experience as a volunteer in a US food bank (Riches, 1986). Indeed, in 1985, John van Hengel and a representative from Second Harvest were advising the first national conference of food banks from across Canada. Held in Edmonton, Alberta in 1989 it was the first step in the establishment of the Canadian Association of Food Banks (CAFB) (Table 3.1).

Canadian food bank directors were however quick to point out 'that advice from their US colleagues (was) not always to be followed' (Riches, 1986), especially 'starting big' and 'going for tax breaks'. In fact many Canadian food banks operated more as food pantries, front line agencies collecting food but handing it out directly to those in need. At the same time some operated as warehouses distributing surplus food to other charities and non-profits. In the first decade this worked sufficiently well, and in the same year that CAFB was founded (Table 3.2), a research study by Professor John Gandy of the Faculty of Social Work at the University of Toronto reported that food banks in Canada had formed a second tier of the Canadian welfare system. In other words charitable food banking was well and truly institutionalized (Gandy and Greschner, 1989).

East to France

Interestingly, it was a Catholic connection between Canada and France (Table 3.2) which led in 1984 to the establishment in Paris of France's first food bank. A nun, 'Sister Cécile Bigot, working with the poor in Paris, had heard about the concept in early 1984 through contact with Francis Lopez' (FEBAc, 2016), an associate of the missionary Oblates, the worldwide Catholic missionary order. Lopez, a native of France had emigrated to Canada and was living in Alberta where he was a founding member of the Edmonton Food Bank. He was well informed about the North American food banking model (OC, 2013). Four months later with the support of charities such as Secours Catholique, Emmaüs and the Salvation Army, the Food Bank of Paris-Ile de France was born (FEBA, 2017).

John van Hengel was also to visit France and Belgium, the first two European countries in which food banks were set up and whose leaders together in 1986 established the European Federation of Food Banks (Table 3.1). Their purpose was to speak with one voice to EU institutions and the multinationals and to support the national development of food banks across Europe. Within a six-year span, Mexico, Spain, Ireland, Italy and Portugal 1992, all Catholic countries, had established food banks and associations with FEBA membership (Table 3.2). Today FEBA operates in 23 countries including the UK, Norway and Denmark with projects in Albania, FYROM-Macedonia, Malta and Slovenia, and with links to Die Tafel in Germany.

South to Mexico

Noticeably, van Hengen's mission also took him south of the US border to Mexico. He had been contacted by Ricardo Bon Echeverria who worked in the food industry for his advice about food banking. In 1987 Echeverria founded Mexico's first food bank in Guadalajara (Table 3.2) and with van Hengel's guidance 'then decided to replicate the (warehouse) model opening other food banks across the country and later a food bank network (Castrejon Violante, 2017b). In 1995 Banco de Alimentos México, the Mexican Association of Food Banks (BAMX) was formed. BAMX was to play an influential role in the establishment ten years later of the Global Foodbanking Network.

What is clear about the early cross border developments in the 1980s (Tables 3.1, 3.2) is how much was owed to American practices and the warehouse model of food banking and to international Catholic outreach including the work of Caritas.

1990s – Gathering pace

Table 3.2 shows the decades and years when the first food banks were established in OECD member states and the formation of national food bank associations.

Expansion in Europe

Food banking continued to gather pace in Europe in the 1990s (Table 3.2). Portugal (1992), Germany (1993), Poland (1994), Greece and Finland (1995) and Austria (1999), with Portugal, Germany and Greece forming national associations and all with the exception of Finland and Germany becoming members of FEBA.

Die Tafel, the German Food Bank Association founded in 1995 and which in 2010 had 850 local food banks chose a partnership model of association with FEBA. The fact that Germany's first food bank was modelled on City Harvest in

TABLE 3.2 Foundation of first food bank / national food bank association, OECD food bank nations, decades and years, 1967–2015

	1960s–1970s	1980s	1990s		2000s		2010s
USA	1967/79						
Canada		1981/89	Portugal	1992/99	Japan	2000/10	Estonia	2010
France		1984/86	Australia	1992/96	UK	2000/04	Iceland	2010
New Zealand		1985	Germany	1993/95	Luxembourg	2001	Suisse	2011
Belgium		1986/86	Poland	1994	Nederland	2002/08	Slovenia	2012
Mexico		1987/95	Greece	1995/2005	Israel	2003/07	Chile	2012
Spain		1987/88	Finland	1995	Turkey	2004	Norway	2014
Ireland		1989/92	Korea	1998/00	Hungary	2005	Sweden	2015
Italy		1989/89	Austria	1999	Slovak R	2005		
					Czech R	2006		
					Denmark	2008		
					Latvia	2009		

Sources: NB Food Banking Inc (1983) International Food Bank Services (1991) Global FoodBanking Network (2006). Food Bank websites/email communication

New York provided further evidence of the spread of American food banking ideas with Lorenz observing that the issue of surplus food was not so much a problem of poverty but one of surplus or waste in affluent societies (Lorenz, 2010).

A Nordic welfare state

In Lutheran Finland, in light of its generally acknowledged securalized and 'cradle to grave' Nordic welfare state, it was surprising to learn in the early 1990s that a 'hunger problem' was brought to public attention by activists working in different relief organizations (Hänninen and Karjalainen, 1994, p.274) with food banks appearing on the scene. Their emergence was a consequence of a deep recession at that time but they have now become 'a fixed feature of the poverty policy landscape' (Silvasti and Tikka, 2015).

There has however been no drive to establish a formal national food bank association. As Anna Salonen has commented whilst food charity is widespread, 'the level of institutionalization within the Finnish food charity system has remained relatively low. Unlike in many other countries, there is no umbrella organization to coordinate food assistance, no national level cooperation with food corporations and no established forms of cooperation with public welfare providers' (Salonen, 2016, p.20). In fact Finnish emergency food relief is heavily dependent on EU MDP food aid (see Ch.3) which is divided between churches and rich religious organizations for distribution. They comprise 'the Evangelical Lutheran Church of Finland, the evangelical Free Church of Finland, the Adventist Church, the Salvation Army and the Street Mission; and other NGOs such as The Mannerheim League for Child Welfare and associations for the unemployed' (Salonen, 2016 p.20; Silvasti and Tikka, 2015).

New Zealand and Australia

In other parts of the OECD world food banking was likewise gaining a foothold. Down Under whilst the first food bank in Aotearoa/New Zealand is reported to have been established at the Auckland City Mission in 1985 (Table 3.2) numbers remained quite small, only developing at a more rapid pace in the 1990s (Mackay, 1994; Uttley, 1997). In fact the term 'food bank' used in Aotearoa refers to the local frontline social agency (Dey, 2017) and as O'Brien has noted 'there is no national organization of food banks although some larger ones supply food to smaller entities' (2014).

Interestingly, however, in 2016 Foodbank New Zealand was incorporated as a publicly funded charitable trust (C. Stephenson email, 18.9.17). FBZ currently offers an information and referral service linking potential applicants with 18 regions across the country and services 169 local food banks, in time planning on-line food banking. The Salvation Army has played a leading role in New Zealand in the development of emergency food assistance through its 60 food banks, and likewise plays an important role in Australia.

Food banks came to Australia in the early 1990s (Booth, 2014, GFN, 2017a) drawing on the ideas of America's Second Harvest. As Lyndal Keevers tells the story food banking as a 'tangible business concept' was brought to Australia by Jeanne Rockey who witnessed food banking in Los Angeles in 1991. She shared this concept with her close friend philanthropist Charles Scarf who in partnership with St Vincent de Paul opened the Sydney Food Bank in 1992. Food Bank Victoria followed in 1993 with Perth in Western Australia and Brisbane in Queensland responding a year later (see Wilson, 1997, pp.38–42). That same year Hazel Hawke, wife of the former Labour Prime Minister Bob Hawke opened the Melbourne Food Bank' (Booth, 2014). I remember the occasion well. I was there and wondering about the politics of food charity for those on the left.

John Wilson writing about the start-up of food banks in the 'Lucky Country' predicated that they would be there to stay (Wilson, 1997). He was right. Over the next two decades food banks became increasingly established across the country. Influenced by America's Second Harvest warehouse model. Foodbank Australia, the national organization, was set up in 1996 with the purpose of providing national policy and organizational focus (O'Hearn, 2004). It was not however until 2010 that Foodbank Australia had administrative and warehouse distributional centres in all six states and the Northern Territory (Keevers, 2010).

East Asia

The first food bank in East Asia was opened in 1998 in South Korea by the Seoul Council on Social Welfare as a response to the IMF financial crisis and a national association was formed two years later to coordinate the development of food banking. In 2012 it had grown to 16 urban food banks, 127 local level food banks and 127 food markets across the country (FBAsia, 2012).

Japan's first and largest food bank was set up in 2000 and was incorporated in 2002 changing it's name to Second Harvest Japan (2HJ) in 2004 (TGG, 2016). In 2010 with the idea of further promoting food banking in Asia core members of 2HJ founded Second Harvest Asia registered as a US 501(c)3 nonprofit organization.

2000–2015 The new millennium

Full membership in the OECD food bank nations club

Membership in the OECD food bank nations club grew at an increased rate in the first decade of the new millennium and by 2015 enrolment was complete (see Table 3.2). Across Europe food banks were being established in short order: the UK in 2000 and Luxembourg, Hungary and the Slovak Republic by 2005 followed by the Czech Republic, Denmark, Latvia by decade's end. Japan, Israel and Turkey were also signing on. Down the final straight from 2010 Estonia, Iceland, Chile, Switzerland, Slovenia and Norway came on board, with Sweden the last to cross the finishing line in 2015.

It is interesting to observe that the first OECD food bank nation club member, the original catalyst, was the USA and the last to join was Sweden, countries at opposite ends of the welfare state spectrum: the first with a residual welfare approach and minimalist role for government, the other built around universality and social democratic beliefs with a central role for public policy. Yet by 2015 all the Nordic welfare states had seen food banks established with Finland leading the pack 20 years earlier.

Also of note is the strong role played by Catholicism in the initial export of US style food banking in the 1980s and early 1990s from North America to Europe and south to Mexico and Latin America. This is not to say that food banks are simply a form of Catholic charitable outreach. That would deny the fact that other Christian denominations such as the Salvation Army, the Lutheran and Anglican churches, the Muslim faith and secular community-based organizations have played significant national and local leadership and organizational roles.

It is likewise important to recognize that in more than half the OECD member states national food bank associations have not been established, though the majority of EU countries are members or partners with FEBA (excepting Finland, Latvia, Iceland, Slovenia and Sweden), whilst other non EU OECD member states have joined the Global Foodbanking Network (Australia, Canada, Chile, Israel, Korea, Mexico, UK, USA). Both FEBA and the GFN are engaged in building the food bank movement from the grassroots to the international level.

It is also worth noting that food banking had been established in more than two-thirds of the OECD countries prior to the Great Recession of 2008, early warning signs of the impact of neo-liberalism and shrinking government support for the poor long before the financial crash. The fact is that the food bank concept has now been embraced in one form or another in all OECD member states adopting either the US style warehouse model or a frontline food pantry way of organizing and delivering surplus food aid. In the meantime food share and meal exhange progammes, food hubs and social supermarkets have blossomed as food safety nets have expanded and become the new normal.

Entrenching food charity safety net

The global spread of US style food banking to the rich OECD world has furthered the institutionalization of food charity and the now taken for granted entrenchment of food safety nets dependent upon wasted and surplus food. From organizational and delivery perspectives what does this mean in practice?

Food banking models

A useful way of addressing this question is considering two models of food banking adopted in the UK. Whilst only a relative late arrival in the food bank nations club, the UK boasts two national food bank organizations: FareShare, a member of FEBA and the Global Foodbanking Network and The Trussell Trust (see Lambie-Mumford, 2017). As Table 3.3 indicates the two organisations illustrate distinctive models of how charitable food banking works in practice and philosophy but with features in common. Both organizations further work to strengthen and expand local and national food safety nets – safety nets of the broken safety net.

The two peak UK food charities have distinct approaches to the national development and institutionalization of collecting, storing and distributing surplus food: the 'top-down' US style warehouse 'food depot' model supplying charities and meal programmes which feed hungry people and the 'bottom-up' frontline food bank 'food pantry' model which is locally sponsored, perhaps with a small food storage capacity, but which distributes its food directly to people in need. Both types will be found in North America and as with any modelling of the social world, they will overlap and vary how they work in practice. One should say they are approximate representations.

Yet at the same time both models have certain features in common: both operate as (private sector) social franchises; both are in the business of sourcing and recycling surplus, wasted or donated food in varying amounts from Big Food, supermarkets or local suppliers and private donors to feed hungry people; both have extensive networks throughout the UK; both depend on national and local food drives; both are heavily dependent on volunteer, or unpaid labour; and both are expanding the boundaries of food safety nets. Their visions, however, reflect different approaches to charitable food banking

Fareshare's origins date back to 1994 when the homeless charity Crisis in partnership with Sainsbury's co-founded CrisisFareshare. Seeking to broaden its scope

TABLE 3.3 Warehouse and frontline 'food bank' models: FareShare and Trussell Trust, UK

Models	Distinct features	Common features
FareShare	Regional FS food depots collecting, redistributing	social franchises
'Warehouse' *food depot function*	surplus food to charities: *indirectly* provides food to vulnerable people	corporate/private sector partnerships recycling surplus food extensive national networks
	Focus: **hunger/food waste**: fresh/longer life food	central brokering with Big Food/supermarkets to access national food supply chain
	Corporate sponsorship	national and joint food drives[i]
Trussell Trust	TT projects: collect, distributes donated food by individuals and local businesses	dependent on volunteer unpaid labour
'Frontline' *food pantry function*	*directly* provides food parcels to hungry people in crisis	EU General Food Law: zero VAT rate imposed on food banks[ii]
	Focus: **hunger/poverty** long life foodLocal church/ community sponsorship	

Sources Annual Report 2016, FareShare UK; Trussell Trust

i   https://www.tesco.com/food-collection/; https://www.trusselltrust.org/get-involved/partner-with-us-old/waitrose

ii *Bio by Deloitte, 2014* see Executive Summary. Comparative Study on EU Member State's legislation and practices on food donation http://www.eesc.europa.eu/?i=portal.en.events-and-activities-eu-food-donations

beyond feeding the homeless, Fareshare was established as an independent charity in 2004. Its social franchise is essentially a private sector partnership between charitable agencies and the food industry which includes over 500 food business comprising supermarkets and retailers and manufactures, growers and packers (see FareShare, 2017). FareShare is a member of both FEBA and GFN.

It works closely with the food industry to eliminate food waste by capturing surplus food and delivering it to charitable organizations which need it, thereby alleviating hunger. FareShare adopts a 'top down' warehousing approach to the collection of surplus food by 20 regional warehouses or food depots across the UK which in turn is delivered by vans driven by volunteers to nearly 6000 partner charities working directly with people in need 'including breakfast clubs, homeless hostels and women's refuges' (see FareShare, 2017).

The Trussell Trust was initially founded by Carol and Paddy Henderson in 1997 as a response to chronic homelessness in Bulgaria where they had both been working as part of the UN Food Programme. However their growing awareness of the extent of 'hidden hunger' in their hometown of Salisbury and the UK led in 2000 to the setting of the UK's first food bank (in a shed and garage) to respond to shocking levels of food deprivation. Four years later the Christian based UK Foodbank Network was established, now more commonly know as The Trussell Trust (TT). It provides emergency food relief to people in crisis. Its mission is to bring 'communities together to end hunger and poverty in the UK by providing compassionate, practical help with dignity whilst challenging injustice' (TT, 2017).

As with Fareshare, the Trussell Trust Foodbank Network (ibid) adopted a private sector social franchise model but working from the 'bottom-up' by partnering with churches and communities to open and operate local food banks, promoting the vision 'that every town should have one' (Lambie-Mumford, 2013; 2017). It also partners with Big Food and the corporate sector. By 2017 423 franchised food banks, which pay a registration fee and thereafter an annual fee to the TT, had been launched across the country. In 2016 it provided 1.1 million 'three day emergency food supplies to people in crisis' (TT, 2017). Food is locally sourced of which over 90 percent is donated by the public at large (TT, 2013). A voucher referral from approved care professionals is required for those seeking access to the food banks which provide three day's emergency food assistance of nutritionally balanced, non-perishable food including advice from TT's trained volunteers for referral to other charities for help in resolving underlying personal crises (see TT website, 2017).

Public Food Aid – MDP and FEAD

As in the USA, the State has all along been an important but often unrecognised player in the development of national food and nutrition safety nets with food supplies – food intervention stocks – being sourced through agricultural surpluses as a way to boost farm income. Food banks have benefited from these supplies. This is also the case in the EU. The European Food Distribution programme for the Most Deprived People (MDP, 1986–2013) 'was launched in the exceptionally cold winter of 1986/87, when the European Community's surplus stocks of food commodities were given to Member State charities for distribution to people in need' (EC-MDP, 2008)'. Notably 1986 was also the year that FEBA was founded with MDP supplies becoming an important source of assured food aid. France and Italy were to become its major beneficiaries.

Given the anticipated depletion of such agricultural surpluses and their future unpredictability the MDP programme was replaced in 2014 by the Euro 3.8 billion Fund for European Aid to the Most Deprived. EU countries were required to contribute at least 15% of the financing of their national programmes which in 2016 amounted to Euro 0.7 billion (FEAD, 2017). Food aid as with other commodities (e.g., clothing, shoes, soap, shampoo) was now available for purchase either by national governments for distribution to partner organizations (eg., food banks and non-profits) or directly by funded partner charities and NGOs to make the purchases themselves.

Two years after the switch to FEAD, at least 14 EU (OECD) countries were receiving either direct food support: Estonia, France, Finland, Greece, Italy, Poland, Slovenia, Spain and the UK; or food in conjunction with basic material assistance: Austria, Belgium, Czech Republic, Hungary, Ireland, Latvia, Slovak Republic; Portugal (EU, 2016). Denmark, Germany, Luxembourg, the Netherlands and Sweden did not apply and Norway and Switzerland as non-EU countries were ineligible. In other words the majority of EU states were willing to prop up their food safety nets by directly partnering with food charities.

Reflections

Since crossing the US/Canada border in the early 1980s and their first flowering in France and Belgium, the national development of charitable food banks is a story of the building and strengthening of American style food safety nets in OECD member states. It has been nearly three decades since food banks imported from the USA became the second tier of the Canadian welfare system, the food safety net of the publicly funded social safety net. Today this is now common place and unremarkable in the OECD food bank nations. There has been little concern as residual charity backed by all sectors of the food industry has led the way.

Governments have played an indifferent but smart game either by feeding food charity with agricultural surpluses, tax exemptions and supportive food waste legislation while simultaneously looking the other way. As the constant demand for lower taxes and austerity driven programs of welfare reform have remained hallmarks of neo-liberal policy making the local growth, institutionalization and international consolidation of food banks have become entrenched in the affluent world.

After all given the early and continuing influence of the Catholic Church and the faith and community based public spiritedness expressed by many thousands of volunteers of all ages and backgrounds in organizing, operating and coordinating the delivery of emergency food aid to people in need what politician would be willing to raise an awkward question or two about human rights and the undermining of public policy.

If so inclined they might have raised their voices about the drip by drip drift away from social security and the income based safety nets of most OECD member states towards the institutionalized food safety nets of the USA. They might also have raised their eyebrows at the steadily increasing global spread of food banking and its corporate capture by Big Food and the world of business all the while waving the banner the banner of corporate social responsibility.

PART II
Corporate capture

4

CORPORATE CAPTURE TO RICH WORLD CONSOLIDATION

The expansion of charitable food banking in North America, Europe and Mexico was due not only to Catholic connections between the early pioneers and the efforts and energy of faith and community-based voluntary organizations. These were certainly early indications of its impending global spread. Yet at root it was a function of the necessary participation of the food and transportation industries which supplied the surplus food and the organizational capacity to make the warehousing model and food distribution system a well-managed business for feeding the poor.

This was not happenstance but attributable to John van Hengel's vision of spreading the food bank word. In 1983 van Hengel left Second Harvest and founded Food Banking Inc which in 1991 became a consultancy – International Food Bank Services. He travelled widely advising on the development of food banking to Canada and Mexico and in Belgium and Spain, and later outside the OECD to many parts of the world (JvH, 2006; Steyaert, 2014). This was to lead over the years to the global corporate capture of food banking, a case of Big Food becoming an integral if not the dominant player in privatized food charity and food safety nets in the rich world resulting in the de-politicization of domestic hunger.

Big Food and corporate partners

While the food banking system has been described as a 'simple' and 'great' idea, if it is to work on the scale necessary to alleviate, let alone end, widespread hunger, it requires stable and ever increasing volumes of surplus food. This is unlikely unless Big Food ensures surplus and unsaleable food moves expeditiously along the food supply chain to food banks for distribution. Attracting the necessary volumes coupled with the logistics of storage, delivery and distribution is challenging to say the least.

As the bedrock source of surplus food it is therefore inevitable that Big Food and its representatives should be sitting at the head table managing or influencing, directly and indirectly, the overall food banking operation. Big Food's vital role is readily apparent in the development and work of Feeding America, the European Federation of Food Banks and the Global Foodbanking Network.

Feeding America

Feeding America, the heir of St Mary's Food Bank Alliance, undoubtedly holds the global blue ribbon award for the largest, wealthiest and most influential national food banking organization. With over 200 affiliated member organizations, with annual total revenue of $2.2 billion in 2015 Feeding America earned third place on the Forbes List of 100 largest US charities (Forbes, 2017), the food charity business could hardly have been better. However has Andy Fisher points out more than $1.9 billion is 'in the form of noncash donations (food and in-kind volunteer time)'. The 'cash budget is only about 5 per cent of its stated budget at $118 million' (Fisher, 2017), a somewhat more modest amount. Still, perhaps little wonder that Feeding America's CEO was earning $606,521 in 2015 (since raised to $651,083) with its top staff being paid between $273, 000 and $377,000 (Fisher, 2017, p.64; Forbes, 2017).

Nor is this level of income restricted to the premier league leading Feeding America. The average CEO salary of eight food bank CEOs across the country was $320,000 with the North Texas Food Bank CEO taking home in excess of half a million dollars in 2013 (see Fisher, 2017, p.64). Perhaps this level of income in the major US food banks is to be expected. After all as we have noted, the American warehouse model of food banking is being emulated in many OECD food bank nations though not, one suspects, at quite such a high rate of remuneration for staff paid to run food charities heavily dependent on voluntary and unpaid labour.

Andy Fisher's timely critique of the US food banking industry in his recently published *Big Hunger. The Unholy Alliance between Corporate America and Anti-Hunger Groups* should be required reading for all who believe food charity is the answer for alleviating domestic hunger and providing a solution to food insecurity. He pays particular attention to the corporate leadership and organizational culture which permeates US food banking.

It is a story of Foodbanks Inc in operation but not confined within US borders. As he writes 'these are sophisticated organizations with multi-million dollar budgets and significant infrastructure, food banks seek highly competent board members with the requisite experience to provide effective governance. Given their close relationships with the food industry, it should come as no surprise that they look to colleagues from these firms to serve on their boards' (Fisher, 2017, p.60). The companies with the leading number of representatives of food bank boards are Kroger and Walmart – grocery/retail; Kraft – food manufacturing; Sodexo – food services and related corporate partners such as Bank of America and Wells Fargo – financial services; Ernst and Young – accounting; Blue Cross – insurance; and UPS – shipping.

While collecting, storing and distributing food is clearly the primary day to day function of the food bank enterprise, acquiring food through donations and fund

BOARD OF DIRECTORS	**(22) Members** including: FOOD BANKS (5) and CON AGRA; KROGER; MONDELEZ; MARS; GENERAL MILLS; P&G; WALMART; PRICE WATERHOUSE, MORGAN STANLEY ...
DONOR HONOR ROLL	**(118) Big Food/Corporate Partners** including:
Visionary *$4 million +/40m lbs food*	(14) CARGILL, CON AGRA, GENERAL MILLS, PEPSICO, WALMART,MORGAN STANLEY ...
Leadership *$1m. +/10m lbs*	(32) UNILEVER, MONSANTO, KRAFT, MONDOLEZ, H-E-B, KELLOGG'S, COSTCO, CAMBELLS, GOOGLE ...
Mission *$500,000 +/5m lbs*	(21) DANONE, DEL MONTE, SYSCO, COCA-COLA, BAYER ...
Guiding *$250,000 +/2.5m lbs*	(21) MARS Inc, PROCTER and GAMBLE, CHEESE FACTORY;STARBUCKS ...
Supporting *$150,000 +/1.5m lbs*	(30) 7-ELEVEN, AMAZON, FORD, IKEA, PRUDENTIAL ...
CELEBRITY SUPPORTERS	**(42) Entertainment Council** including Ben Affleck, Courteney Cox, Sheryl Crow, Matt Mamon, Josh Groban, Karolina Kurkova, Kimberly Williams-Paisley ...

FIGURE 4.1 Board of directors, selected Big Food/corporate partners by donor honor roll and celebrities, Feeding America (2014–15)
Source: FA, 2015, FA, 2017d,e

raising is paramount. This is reflected in Figure 4.1 which illustrates Feeding America's business acumen and philanthropic power. It includes 117 corporate partners and donors drawn from the Big Food conglomerates and leading global corporations to say nothing of the star power of a 139 celebrities, movie stars, TV personalities and musical entertainers to support its cause.

While the Entertainment Council is designed to leverage celebrity support, Feeding America also receives also financial donations between $5000 and $149,999 from hundreds of organizations and individuals; food and grocery donations from right across the food industry. Notably the van Hengel Society honours individuals leaving legacies and bequests. It is clear that Feeding America is well established to collect and distribute surplus food.

It is however not surprising that given the history of the troubling paradox of hunger amongst plenty in the USA and its strong belief in the food safety net as the primary response to hunger and poverty that Feeding America's 22-member board of directors has significant representation from Big Food and the corporate elite. These are after all understood to be 'the' food experts with the corporate values, knowledge and experience to feed the nation, indeed the world, and get things done. Food aid or emergency assistance which draws on supplies of surplus food is simply factored into the industrial food supply chain and is part of everyday business. Moreover one of Feeding America's important functions is certifying member food banks.

However, as Fisher observes the corporate culture which over time has taken over US food banking has spawned those astonishingly high CEO salaries which serve as 'indicators of the degree to which food banks have become "mainstream, respectable and rich"' (Fisher, 2017, p.23). As he says, it is all about 'feeding the need'. The focus is on the outputs: acquiring ever increasing volumes of surplus food, sorting and delivering it as quickly as possible and reporting the numbers of meals served and people fed. The causes and consequences of food insecurity are not high on the agenda.

Whilst food drives and fundraising are built around themes of 'ending' or 'alleviating' hunger, the strategies of corporate food banking are a long way removed from the goals of food and social justice and from advocating for a living wage let alone adequate welfare benefits. The fact is that mainstream food banks have become dependent upon the corporate good will of the industrial food system. Feeding America has been a powerful catalyst for the corporate capture and national consolidation of charitable food banking in the United States an idea which has been emulated and acted upon within other OECD food bank nations.

European Federation of Food Banks

The European Federation of Food Banks (FEBA) grew from a joint partnership between the Paris and Brussels food banks in 1986 to an international non-profit organization which by 2017 with members from 23 European (and OECD) countries representing either national federations of food banks or a single food bank (FEBA, 2017a). FEBA has played a significant role supporting the development of food banks across the EU as well as supporting current projects in Albania, Macedonia, Malta and Slovenia and with sustaining links to Die Tafel in Germany. An original commitment was to speak with one voice to the EU for the purposes of acquiring surplus food from the then MDP programme, reorganized as FEAD in 2014 (FEBA, 2016c).

Since its foundation FEBA has worked closely with international charities such as Caritas, the Salvation Army, the Red Cross and with social centres and local charities. In 2015, member food banks in 23 EU countries partnered with 33,2000 charitable organizations distributing enough surplus food to provide 2.9 million meals (531,000 tons of food) to 5.7 million people. Of the 15,500 people who are currently active in operating the European food bank project, 90% are volunteers (FEBA, 2016a). There can be little doubt that community and faith-based charities are key players in building and consolidating food safety nets across Europe, a process that has been further facilitated by growing public support for addressing the environmental food waste crisis and the idea of the circular economy (FEBA, 2017b).

FEBA's governance structure – General Assembly, Board of Directors and Team of 'regular volunteers' – is of interest, differing markedly from that of Feeding America. Its 12-person Board has no active Big Food representatives from. It is elected from the 40-member General Assembly which represents the constituent food bank associations (FEBA, 2017c). FEBA's Team therefore is of note, currently comprising 12 'regular' volunteers who in the past have experience of business, public services and different associations (FEBA, 2017d). Their responsibilities include managing the Federation including building relationships with government and the private world of business.

One member of the FEBA Team has designated responsibility for forging and strengthening EU relations. FEBA is not only committed to acquiring and distributing surplus food from FEAD to ensure the hungry have access to nutritious meals but also to helping achieve the EU's 2020 goals of enhancing social inclusion goals and reducing the numbers of those experiencing poverty by 20 million by 2020. This includes those suffering from monetary poverty, severe material deprivation and very low work intensity, the jobless and precariously employed.

However, while differing in organizational structure, there is a mirror operational reflection of Feeding America. Three members of the FEBA Team have designated responsibility building and strengthening private partnerships with the food and non-food industries. As with food bank associations in all OECD member states they share a necessary dependence and strategic focus on the world of Big Food and the corporate sector (FEBA, 2017d). As Box 4.1 shows the business of finding, funding and distributing sufficient volumes of surplus and wasted food to feed millions of hungry people across Europe depends for the most part on the engagement, support and corporate social responsibility of the principal actors within the food system: Big Food multinationals and local food producers and retailers, supermarkets, restaurants, and corporate partners in the finance, transportation, equipment and software industries.

BOX 4.1 BIG FOOD MULTINATIONALS, CORPORATE PARTNERS AND FOUNDATIONS, FEBA, 2015

Big Food Multinationals: Unilever, Cargill (founders); Bonduelle, Danone, Kellogg's, Mondelez International, McCain, Nestlé, Tesco

Corporate partners/foundations: Metro (international wholesale); AXA Assistance (global insurance); Carrefour (multinational retailer) Foundation; LiFT (global software/insurance); Hitachi (conglomerate); Brambles (global/pallets, crates, containers – CHEP, IFCO); Bloomberg Philanthropies/Bloomberg LP (global software, data and media); Auchan (global retail, hypermarket/supermarkets), Société Générale (multinational banking/financial services)

Source: FEBA, 2017d

Charitable food banks whatever their size and scope are clearly dependent on voluntary labour but they need to be backstopped by the Big Food brands. Equally as Figure 4.1 indicates they also require the support of corporate partners which can assist either with funding and food drives; pro bono legal advice and accounting; equipping warehouses with freezes, pallets, containers, forklift trucks; vans, trucks and rail cars for distributing food; software technology to assist with organizational management and the many factors which are required to run a successful food bank organization (FEBA, 2017e).

Judging by its website, reducing food waste is as much a priority for FEBA as reducing hunger and food poverty. For that reason access to businesses up and down the food chain is vital to its work of acquiring surplus food. Obviously if food waste is understood as a solution to food poverty you need the relevant corporations on your side.

What better than the partnership signed in June 2016 between FEBA, Food-DrinkEurope (FDEU) and EuroCommerce (EUComm). FDEU is a more than three-decades-old international confederation including 26 national food-related federations and 19 major food and drink companies (FDE, 2017a), while EUComm represents 'national federations and companies in the retail, wholesale and international trade from 31 European countries … linking 5.5 million companies most of which are small or medium sized' (EComm, 2016). In 2016 the three federations jointly launched a new guide entitled *Every Meal Matters* designed to encourage and make it easier for food manufacturers and retailers to donate their food surplus to food banks (FDE, 2017b).

FEBA's partnership with these two powerful transnational food federations should prove beneficial. They both support the UN Sustainable Development Goal 12.3 which specifically targets the halving of global food waste by 2030 (see WRI, 2017). In 2015, at the urging of the Netherlands government, Champions 12.3 was formed comprising a coalition of executives from governments, business, international organizations, research institutions, farmer groups and civil society to inspire ambition dedicated to achieving the 2030 goal (CH12.3, 2017).

Despite its membership-based organizational structure, FEBA it would appear has also played its part in facilitating the corporate capture of food banking across Europe. Certainly it acts as a force for the development and consolidation of food banking as a response to poverty. Yet its commitment to fighting food waste brings with it an inevitable dependence on Big Food and its corporate partners as the primary source of surplus food for reducing hunger and poverty. FEBA's reliance on the corporate food sector is shared by its international partner the Global Foodbanking Network based in Chicago, Illinois.

Global Foodbanking Network

The international corporate capture of food charity was further advanced in 2006 when International Food Bank Services consulting organization became the Global Foodbanking Network. A decade later it operates in 34 countries worldwide, partnering with FEBA and supporting a regional network in the Middle East and North Africa (GFN, 2017a). GFN's membership includes food bank association and organizations from nine OECD member states: Australia, Canada, Chile, Israel, Mexico, South Korea, Turkey, the UK and the USA and indirectly those European food banks affiliated with FEBA.

GFN's founding resulted from discussions in 2005 between John van Hengel, Bill Rudnick, the late Bob Forney and Chris Rebstock, all from America's Second

Harvest board of directors, and Ricardo Bon Echeverria (the President of BAMX) and representatives from the national food bank associations in Argentina and Canada. Prompted by the development of food banking in Argentina and the increasing numbers of overseas requests being received by the North American food bank associations for advice and interest in certification, the idea was to form a global network of food banks (GFN, 2015). Van Hengel approached Bob Forney of America's Second Harvest asking him to take on this task. At a meeting in August that year in Mexico the decision was made. Later there was an event at Los Pinos, Mexico's official presidential residence, with President Vicente Fox Quezada to acknowledge the participation of Mexico and the creation of the Global Food-banking Network (Castrejon-Violante, 2017b).

The GFN was established as 'an international nonprofit organization that fights world hunger by creating, supporting and strengthening food banks around the world, in countries outside the US' (GFN, 2017a). Its founding reflects the corporatization of North American food banking and its promotion with the support of Big Food. Robert Forney, the GFN's inaugural President and CEO who had previously held the same position at America's Second Harvest, and prior to that as CEO of the Chicago Stock Exchange, was explicit. He declared GFN's mission was 'to promote and support food banking worldwide' and it was to be informed by 'the idea of food banking as a *business* [my italics] solution to hunger' (GFN, 2006). In other words the GFN was to serve as the export agency for corporate America's brand of food banking.

This was reflected in GFN's first board of directors. It comprised executives from the national food bank associations of Argentina, Canada, Israel, Mexico, the UK and USA and also representatives from the corporate world of food retail, transportation and law: Sodexho Inc (USA), Kellogg Company (USA), Unilever-Mexico; Dot Foods Inc (USA); DLA Piper (USA). Big Food donors also included: Cargill, H-E-B, Kraft and the P&G Foundation (GFN, 2007).

GFN's first Newsletter (2016a, 2016b) was clear about the corporate message. It describes the newly created international NGO being well-situated between the world of food banking and corporate expertise and with the influence to bring about desired results. 'Leveraging' was named as a key tool at its disposal: to build on the experiences and efficiencies of food banking in different countries by strengthening existing ones and building new ones; to facilitate access to multinational organizations and furthering productive relationships between them and food banks and their networks; and, it should be noted, to build support for the 'development of public policies and programmes by national governments in support of food banks' (GFN, 2006).

In particular GFN argued it was well-situated to attract the resources of the global food and grocery industry; provide a business solution to the problem of surplus food; and strengthen social service agencies across the world by helping them to secure food at low cost (GFN, 2006). In other words a win–win for all. It made such claims with a high degree of assurance. After all leveraging corporate links with Big Food multinationals, international finance and law to say nothing of celebrities had for many years been driving the work of America's Second Harvest

(A2H), later rebranded as Feeding America. In 2006 A2H's leading corporate partners included ConAgra Foods, Kraft, Kroger, Supervalue and Wal-Mart (A2H, 2006) with donations ranging from one to ten million dollars.

In light of A2H's CEO and President moving to lead the Global Food Banking Network it is no surprise that by the time of its launch in 2006 multinational partners and donors had already signed up to support food banking's global expansion. Corporate donations were received from Cargill, Kraft Foods, Procter and Gamble, and the international law firm DLA Piper which provided pro bono legal services and office space in Chicago where its international headquarters remain today (GFN, 2006).

The process of global consolidation gathered pace over the past decade. Today, GFN's 13-member board has a different mix of expertise with the majority of directors representing national food bank associations including five from overseas. Feeding America still has a place at the table and the position of General Secretary is held by FEBA. (GFN, 2017b). Nevertheless as Box 4.2 shows following the leadershp of Feeding America the pattern of GFN's Big Food and corporate partnerships has shown significant expansion worldwide.

BOX 4.2 BIG FOOD MULTINATIONALS AND CORPORATE PARTNERS (EXAMPLES), TYPE OF ORGANIZATION AND COUNTRY, GLOBAL FOODBANKING NETWORK, 2013

Big Food multinationals

USA: Campbell Soup Company, Cargill Inc, ConAgra Foods, General Mills – General Mills Foundation; Griffiths Laboratories Foundation, Inc; H-E-B; H. J. Heinz – H. J. Heinz Foundation; Ingredion, Inc; The Kellogg Company – Kellogg's Corporate Citizenship Fund; Mondeléz International Foundation (formerly Kraft Foods Foundation); PepsiCo, Inc; The Coca-Cola Company; Walmart (ASDA in UK);

Canada: McCain Foods; *UK/Netherlands*: Unilever; *Switzerland:* Nestlé; *France*: Carrefour – Foundation Carrefour, Danone, Sodexo

Corporate partners:

Anti-hunger Non-Profits: Feed My Starving Children; MAZON: A Jewish Response to Hunger; Share Our Strength. *Companies*: Black & Veatch; Hilton Worldwide; Informatica Corporation; *Financial and real estate institutions*: Bank of America Charitable Foundation; BBVA Compass; BNY Mellon; Jones Lang LaSalle Charities *Foundations*: Abbott Fund; Caterpillar Foundation. *Law:* DLA Piper LLC (US) and DLA Piper Foundation; *Professional sport*: Taste of NFL; *Service Clubs*: Lions Clubs International, Rotary International; UNFAO (17)

Derived from GFN, 2013a

Big Food is the dominant player in the GFN but as with FEBA it is also clear that multinational companies with business experience and expertise in corporate finance, law, transportation, logistics etc are also vital to the success of collecting, storing and distributing domestic food aid (GFN 2017c). These forms of board membership, partnership, dependence and corporate branding are reflected to greater or lesser degree in the national, state and provincial leadership of food banking within the OECD. The business of food charity is not simply promoted by the day to day activities and experiences of food banks themselves but by purposeful education and training.

Food Bank Leadership Institute

In 2007 the year following its foundation the GFN took another step forward when, in partnership with H-E-B ('Here Everything's Better') a privately owned grocery company operating an extensive supermarket chain in Texas with subsidiaries in Mexico, it established the Food Bank Leadership Institute (H-E-B, 2017; GFN, 2007).

Based in Houston, Texas, the FBLI through its annual institutes provides 'global forum for education, technical training, and best practice sharing for those involved in food banking'. Its intent is to improve food banking's efficiency and effectiveness informed by sound management practices recycling ever increasing volumes of surplus food to feed the hungry, backed by Big Food corporate know-how. FBLI's prime purpose is to attract, educate and train food bank leaders and social entrepreneurs from across the globe whose primary commitment is to 'fight hunger and food waste through food banking', the core of the GFN message (GFN, 2016a; FBLI, 2015). Little wonder the public perceives food banking as the answer to both food poverty and food waste.

Unsurprisingly FBLI's corporate and foundation partners include those who support Feeding America and the Global Foodbanking Network (GFN, 2006) including its founding partners Cargill, General Mills, Kellogg's and DLA Piper. Since 2007, FBLI's annual institutes have been attended by nearly 400 individuals from 55 countries. In 2015 these included 75 food bank leaders from 35 countries among whom a third of those participating were from OECD member states: Australia, Canada, Chile, France, Hungary, Israel, Italy, Mexico, Poland, Turkey, the UK; and from FEBA and Feeding America. Interestingly speakers at that year's Institute also included high ranking officials from the UN FAO and World Food Programme and suggesting perhaps that such UN participation affirms the value of food banking (FBLI, 2015).

In spreading the corporate food banking message, FBLI engages a two-way process. Not only does it provide global education and training for food bank leaders but an equally important goal is ensuring that those working in the food industry participate in its education and training to learn the importance of corporate social responsibility (CSR). While there are many definitions of CSR the *Financial Times* understands it as a movement for 'encouraging companies to be more aware of the impact of their business on the rest of society, including their own stakeholders and the environment' and to adopt 'business approaches that contribute to sustainable development by delivering economic, social and environmental benefits for all stakeholders' (FT, 2017).

The perspective that food waste reduction is an answer to domestic hunger well informs the joint GFN and FBLI's corporate task of expanding food bank networks locally and globally. Given Big Food's continuing support as well as that of other powerful non-food corporate actors such as finance, transportation and law it is not difficult to see how food charity and food safety nets are becoming increasingly perceived as common sense and publicly acceptable effective responses to food insecurity and domestic hunger.

However, the corporate capture of food charity is not limited to the world of Big Food and its business partners turning the issue of hunger into a matter for the private sector and charity. Other economic and social forces are at work within popular culture rendering charitable food banking a normal and customary activity in the affluent 21st-century first world.

Bread and circuses

Food banking in the US likewise benefits from the corporate good will and star power of Hollywood and Nashville celebrities as well as the professional sports machine. They serve to remind the millions of fans of the hunger issue promoting and keeping alive the legitimacy of charity and doing one's bit.

Celebrity star power

Feeding America's citation of 139 celebrities (FA, 2017d) who contribute to food banking across the USA are key players in promoting food banking and strengthening the charitable food safety net. They are household names from film, TV and the music industry including such stars as Alan Alda, Emily Blunt, Annie Lennox, Beyonce, Ellen Degeneres, George Clooney, Jennifer Anniston, Elvis Costello, Aretha Franklin, Bob Dylan and Sarah McLachlan. These are a few of the names known to me. The lengthy celebrity donors' list is not a shabby one.

Nor is such fundraising and support left to chance. Feeding America has an established and well-organized Entertainment Council whose members 'actively mobilize the public in support of Feeding America via media relations opportunities, cause marketing programs and public policy initiatives' (FA, 2017e). Sitting on the 44-member Council are celebrities such as Academy Award winning actors Ben Affleck and Matt Damon; Tony and BAFTA winner Scarlett Johansson; Actor/producer David Arquette; actress/director Courteney Cox; sporting starts Jason Grilli, major league pitcher, figure skating Olympic Gold medalist Scott Hamilton; singer and song writer Josh Groban; and former supermodel Karolina Kurkova. There is an expectation that celebrities will publicly voice their concerns about domestic hunger in America and also participate in local volunteer activites (FA, 2017e).

In Canada the music industry similarly plays an important role in promoting food charity. A Google search quickly uncovers the names of rock stars and bands such as David Myles, Justin Bieber, the Elastic Band, Rush, The Tragically Hip, BareNaked Ladies, Sarah McLachlan and Neil Young who have been associated over the years with food bank appeals and donated their concert or album proceeds to food charities (CBC, 2010). Why not if celebrity appeals to community compassion and resonates with music lovers, young and old?

Professional sports

The professional sports industry also plays a significant role in North America as a site for charitable giving and feeding the hungry. In the US dating back to 1992 the 'Taste of the NFL' has been raising money for food banks with the support of the country's top chefs and National Football Leagues's star players. It is institutionalized and rolls out every year. Annually a Kick Hunger Challenge is held between competing NFL Teams to raise funds. In the 2016–17 season the top 10 of the 32 teams competing in the challenge raised over $440,000 enough for 3.2 million meals (KHC, 2017).

In Canada, on a lesser scale, it is a similar story on game days in the Canadian Football League. The Canadian Television Network (CTV) has for many years been broadcasting food bank drives through Purolators 'Tackle Hunger' programme. The links to food banking are transparent. In the mid 2000s when the CAFB was reorganizing its management structure and rebranding itself as Food Banks Canada, its treasurer was a representative of Purolator Courier Ltd. Likewise in ice hockey-mad Canada NHL and WHL teams support food bank drives, make financial donations, host dinners for the poor, and organize gift programmes for needy children, especially at Christmas time.

Food banks' dependence on such financial support to ensure steady supplies of food is critical. In 2012 the Victoria-based *Times Colonist* reported that the months-long National Hockey League lockout by the Board of Governors (the franchise owners) in 2012 was not just frustrating for hockey fans – it certainly was, I remember well – but even 'penalized some food banks' as 'food drives usually conducted during the holiday seasons by teams and sports bars' were put on hold. Food banks in Ottawa and in Montreal, home of the famed Canadiens, saw food donations decline (TC, 2012). Interestingly the just as famous Toronto Maple Leafs did not respond to a request for comment. However in 2017 the Lady Jays working on behalf of the Major League Baseball Toronto Blue Jays once again compensated by stepping up to the plate with organizing their 33rd successful annual food bank drive.

These organized and ritualistic acts of star power philanthropy connecting fans with the hunger question is an important source of revenue for the food bank industry in North America. Unlike in Roman times it is the business world not the government offering free bread and circuses. Celebrity star power through the corporate world of professional sports and entertainment plays an influential role in developing and sustaining charitable food banks and food safety nets all the while helping to divert the public and political gaze from the systemic causes of domestic hunger.

Whether this reflects an effective role for corporate social responsibility or is a case of bread and circuses functioning to distract both performers and the public from the underlying causes of hunger and poverty and an indifferent state is perhaps a moot point. The fact is it legitimizes and further entrenches food banks as popular and deserving causes.

Mixed media messages

The spotlight must be shone on power and influence of the media and the varying ways and in different times it has either highlighted or ignored the problem of hunger and poverty; been indifferent or inconsistent in investigating its causes; and supported or shown hostility to those begging for food. For the most part, it appears, the mainstream media has uncritically facilitated the growth and privatization of domestic hunger whilst neglecting, deliberately or not, the role of government and public policy. Critics writing about the media and food banking make a number of concrete points about the ways in which it safeguards and promotes food charity.

Perhaps Joel Berg, activist, author and CEO of Hunger Free America asks the crucial question: why was it in the USA that in the post 1960s when the domestic hunger and poverty problem was worsening that the issue was rarely covered by the national media; and, on those occasions when it did, why the media no longer assumed it was government's responsibility to solve the problem. Looking back at those times Berg writes 'the media's direct or implicit question was "how can a country this wealthy let children go hungry?"' – a much needed question for the media to raise in today's OECD food bank nations.

Blaming the victim

Then in the 1980s with the Reagan era ushered in and the Clinton and Bush years which followed, Berg notes the media's implicit question became 'Why are all those undeserving people getting benefits with our tax dollars?' (Berg, 2008). Clearly the steady diet of neo-liberalism and punitive welfare reform had successfully changed the public conversation. A disinterested media neglected the role of public policy enabling corporatized food charity to flourish.

Berg's comments were to echo across the Atlantic. In the UK where food banking spiked dramatically following the 2008 Great Recession, Rebecca Wells and Martin Caraher of the Centre for Food Studies at City University have written about the UK's tabloid media framing those using food banks as members of the undeserving poor (Wells and Caraher, 2014). Separating people in poverty as 'deserving' or 'undeserving' was very much a case of pulling out old stereotypes of victim blaming enriched by accusations of 'scroungers' living off free food while shirking employment opportunities (Wells and Caraher, 2014). Such coverage certainly diverts attention from questioning government austerity driven sanctions policies.

Kayleigh Garthwaite, research fellow in social policy and author, confirms this with her insightful study of life inside food bank Britain. She comments that 'the powerful political, media and public discourse continues to question the lifestyles and attitudes of people using the food banks'. As one of her food bank respondents reflected 'You see it in the papers, don't you?' (Garthwaite, 2016) as well as in the spate of UK media-led reality TV shows depicting life on benefits which she pointedly describes as 'poverty porn'.

The charity fixation

However it cannot be argued that the mainstream media, what Wells and Caraher (2014) refer to as the broadsheet (serious) media, has been opposed to charitable food banking. Far from it. The reasons are not hard to find. Berg in the USA points to the 'extreme consolidation of ownership of local and national news media by a handful of giant corporations'; of many well-paid journalists being out of touch with working class life; and of the media no longer maintaining a poverty beat. Significantly he underlines the media's fixation on charity and 'feel good' or 'feel bad' stories while rarely considering the role of government (Berg, 2008).

In light of such corporate media concentration it is not difficult to believe that for the most part the mainstream media is persuaded that domestic hunger, if in fact it does exist (essentially a non story as really a matter of famine in the South), is best left to charity working hand in glove with Big Food and the multinationals fuelled by corporate social responsibility. Yet as author David Rieff has warned we should beware the business media's fascination with innovators and the view that 'the Golden Age of zero hunger and the end of poverty that philanthrocapitialism stands on the threshold of ushering in' could be realized 'if only the politicians would get out of the way' (Rieff, 2015, p.241).

In Canada with its long history of food banking, I commented over thirty years ago on the power of the media, specifically the CBC, Canada's national public broadcaster. The role it has played and continues to play in supporting the food bank industry has long been apparent. In the early days the CBC and independent television and radio networks began by airing shows supporting food banks. Staff organized food bank appeals in local shopping malls and have even visited schools asking for donations. Mark Cardiff, a CBC producer in Halifax, explained the public broadcaster's support for a benefit concert for a Cape Breton food bank: 'We want to draw attention to the situation these people are in. We want to raise some money for the food bank and show that the CBC does care about its broadcast community. And, not least, it's to provide ninety minutes of first class entertainment' (Riches, 1986).

Dating back three decades the practice of the CBC hosting annual food bank appeals has been celebrated around Christmas time as a ritual national food charity event. Rewards for donating food or money include studio tours, meeting TV/radio personalities and incentive prizes including holidays for two at a luxury 'spa' hotel (CBC, 2015). In more recent years as concerns about food waste have grown, the food bank argument that reducing food waste is the answer to food poverty has been fed by the media's uncritical and inconsistent reporting.

When a compliant public broadcaster becomes the news by raising funds and actively supporting the message that food banks are the answer to domestic hunger, it is little wonder that food charity has become entrenched. In this respect the actions of a compliant public broadcaster are not dissimilar to Feeding America's embrace of corporate philanthropy south of the border. Awkward questions about human rights, the role of government and public policy are neglected as domestic hunger is socially constructed as a matter for corporate food charity.

In Lutheran Nordic Finland where charitable food banking is not meant to be, the media response according to Tiina Silvasti and Jouko Karjalainen has been contradictory. There is, they write 'a discrepancy, with charitable food aid presented as an illegitimate form of social security for the welfare Nordic welfare state but, at the same time legitimate for the churches' (Silvasti and Karjalainen, 2014). This inconsistent interpretation as both food security experts argue is now well established in the Finnish media and hardly ever challenged. The problem is that it eats away and 'weakens the understanding of food insecurity as a political question' and allows governments to 'silently' avoid their constitutional responsibility to address domestic hunger (ibid). Food charity has become mainstream.

A political question

In Catholic Spain, where food banks are now organized in each of the country's 55 provinces, Karlos Pérez de Armiño, international relations professor and research director, has written they 'acquired a notable presence in the mass media' at the height of the economic recession (Pérez de Armiño, 2014) but less so lately as times have improved. Significantly he comments that 'El País (the most read newspaper; social-democratic orientation) published several articles containing references to food banks' experiences as a way of solidarity from society itself, but in the framework of wider information related to the social impact of the crisis (increase of malnutrition and poverty) and of the lack of public policies to face them (KPdeA, email communication 19.8.17), in other words a political matter.

Fortunately, the UK's *Guardian* newspaper likewise understands domestic hunger as a political question and the solution being central to the role of government and public policy. While recognizing the dilemmas posed by food charity for politicians of the right and left, its social policy editor Patrick Butler has recently asked how long the British public 'will tolerate charity food as a substitute for more efficient ways of tackling poverty and food insecurity' (Butler, 2017). Yet such clarity in the media is rarely found even where you would expect it.

Corporate capture: hunger as a charitable business

The high income world's corporate capture of food charity, perhaps best understood in private enterprise terms as its transformation to Foodbanks Inc has over time led to the social construction of domestic hunger as a matter for the business of charity and not the state and public policy. The backdrop has been more than 35 years of a steady diet of neoliberalism: economic growth which has failed to raise all boats; recessions whose damaging effects linger on; restructuring labour markets with an increasing mix of precarious part-time work and stagnant wages; regressive taxes and income distribution; shrinking governments, austerity politics, privatization, welfare reform and broken social safety nets. The consequent surge of corporate food charity announces the de-politicization of hunger and poverty.

Looking back at the USA with its tradition of public food assistance programmes and food safety nets, the inspiration that sparked the first charitable food bank as a local response to the rediscovery of hunger in 1967 is no surprise. The warehouse model of collecting, storing and distributing wasted and surplus food was indeed a simple and publicly accepted idea which with powerful religious, community and corporate backing has expanded worldwide, including within the rich OECD.

Different forces have been at work extending the global reach of the food bank movement, steadily building from local origins and small beginnings in the USA to national associations and global networks. The appeal to feed hungry people has resonated with concerned citizens as a moral imperative requiring a practical and immediate response. As John van Hengel would often quote ' the poor we shall have with us always, but why the hungry?'

The response depended in the first instance on individual, community and faith-based commitments and by the support of religious charities such as Caritas, St Vincent de Paul, the Salvation Army, the Anglican Church and even the Lutheran Church. Community organizing skills leading to the development of different food banking models; connections within and across territorial boundaries and the willingness of thousands of volunteers have been essential to its success. It has been about people doing their bit. It is little wonder that food banking is widely perceived as an effective community response and that volunteering has come to enjoy such strong public legitimacy.

Austerity politics, inadequate wages and welfare benefits have certainly driven the demand for charitable food relief. However the availability of wasted and surplus food has been a prime factor in the building of the corporate food bank model. Government procurement of agricultural surpluses as in the US or EU as well as grants, food safety legislation and tax breaks in different countries have played an important role as has the support of the food retail sector. Without the participation of supermarkets, grocery stores and food services such as restaurants and catering companies, food banks would have a limited shelf life, as most if not all are frequently running out of food. Food drives therefore play a significant role in encouraging donations from the public with the goal of stocking the food bank shelves unless they are in a position to purchase food.

Yet the truth of the matter is that food banking today would not be the entrenched institution it has become in OECD member states without the business acumen and powerful corporate backing of Big Food and supportive transnational players in the finance, transportation, equipment, software and logistics industries. Their private wealth and sense of corporate social responsibility along with that of North American star power celebrity in professional sports and the world of entertainment has brought about the corporate capture of charitable food banking. These partnerships with the transnational corporate world are readily transparent though as Fisher argues they have created an unholy alliance between Corporate America and anti-hunger groups (Fisher, 2017).

It therefore should not go unnoticed that Feeding America and Big Food's Global Foodbanking Network has its own Food Bank Leadership Institute for promoting food banking as a worldwide business and advocating that public policies support their further development. By offering global training and food bank certification, its goal is to further export, grow and consolidate the US 'warehousing' model of food assistance and safety nets to low income countries while securing its reputation to the rich world.

Meanwhile the mass media is largely disinterested or accepting that corporately backed charitable food banking rather than public policy is the effective response to food poverty. It thus becomes difficult publicly to debate GFN's seemingly common sense message, particularly in times of austerity, that 'food banks are the link between food waste and hunger'.

Reflections

Such powerful forces manufacture the public belief that food poverty is best left to the corporate social responsibility demonstrated by food banking, inc but what Tim Lang has called 'crude and inefficient charity' (2015). Institutionalized food charity and its corporate capture inevitably depoliticizes hunger allowing the indifferent state bent on ever lower taxes to hold its neoliberal course and look the other way. Public policy is neglected. Should we not be concerned that the State now relies on Big Food and its corporate partners in alliance with charitable food banks, the charity economy, to redistribute wasted and surplus food to feed hungry people thereby entrenching food assistance rather than income-based social security as effective responses to domestic hunger and poverty.

Food waste for hungry people has now become accepted as compassionate, practical and common sense: a simple solution, even a 'great' idea to resolve two complex issues. It suggests we – the State, the private sector and civil society – are all doing our bit. Yet are we? Perhaps it is first necessary to consider the Global Foodbanking Network's claim that 'food banking serves as the link between food waste and hunger' (GFN, 2011) and that surplus food distribution is the way to solve the structural problems of both food poverty and food waste.

5

CORPORATE FOOD WASTE

Manufacturing surplus food

Food waste: the imperative to act

The crisis of global waste, including food loss and waste, and how to reduce it is the subject of a growing collection of UN reports, national and international think tank studies, academic research and media articles. In particular the 2015 UN Sustainable Development Goals (SDGs) contain strong messages for the rich world about climate change, greenhouse gas emissions and saving planet earth.

The SDGs stress the need for governments and global institutions to exercise responsibility in ensuring 'sustainable consumption and production patterns' and declaring that 'achieving economic growth and sustainable development requires that we urgently reduce our ecological footprint by changing the way we produce and consume goods and resources' (SDGs, 2015b, Goal 12). By 2030 the goal is to halve per capita global food waste at the retail and consumer levels and to reduce food losses along production and supply chains, including post-harvest losses.

The message is directed at the food industry, its partners and consumers to play their part in recycling and reducing waste, and that the first world must take ownership and show leadership. There is an imperative to act given the global epidemic of food loss and waste as symptomatic of undeniable environmental and food system issues and the need to respond to domestic hunger in the rich world. Little wonder that food waste has become a powerful driver of surplus food redistribution with Big Food's championing of charitable food banking as a frontline strategy for reducing both food waste and food poverty, two issues seemingly joined at the hip.

Food waste and food poverty

Notably the first guiding principle of the Charter of European Food Banks is 'to fight against wastage'. Whilst recognizing the necessity to ensure food safety the task of food banks in stocking their shelves is 'to recuperate agricultural surpluses; excess production in the food-processing sector, or goods which are non marketable but nevertheless consumable, and surplus food from canteens or restaurant chains' (CEFB, 2017). In tandem the Global Foodbanking Network advances its message that food banks are 'the link between waste and hunger' (GFN, 2013b). Surplus food redistribution promises to solve the paradox of widespread domestic hunger and reduce mountains of wasted food in the OECD world. Yet how can this be?

Food waste and domestic hunger are two critical but separate structural issues, the former a symptom of a dysfunctional global food system and the latter a consequence of income poverty and inequality, broken social safety nets, pro-rich income redistribution and neglected human rights. Neither are solutions to the other. Distinct public policies are required to address these systemic problems. The neoliberal reliance on surplus or 'left-over' food can never be more than a palliative for 'left-behind' people.

Thinking about food waste food and surplus food

The troubling question of the food waste–surplus food argument is whether diverting increasing volumes of unsellable surplus food to supermarket shelves really is a 'win-win' both for reducing food poverty as well as lowering methane producing carbon pollution generated by food waste otherwise rotting in landfill. Many influential actors – the anti-hunger movement, Big Food, environmental agencies and governments – support this case. It has certainly driven the rise of food bank nations: surplus food redistribution as the antidote to domestic hunger and food waste in the rich world.

A pot pourri of terms

In thinking about food waste diverted from the from the landfill to the food bank I have found it useful to note a *pot pourri* of commonly used terms which are used to explain and support this argument:

> **Food waste**: food loss; wasted food; food leftovers; inedible food; unusable food; expired food; food scraps; unmarketable food; squandered food; thrown away food; discarded food; dumpster diving; binned food; uneaten food/not consumed: … **to the landfill**
>
> **Surplus food**: gleanings; excess food; spare food; additional food; food leftovers; unused food; food recovery; unsellable food; rescued food; food salvage; recovered food; donated food; recycled food; food abundance; uneaten food for distribution; edible food; food aid … **to the food bank**

In different ways they give meaning and substance to the serious nature of domestic, local and global food waste; and, by reflecting and appealing to mainstream opinion, and socially constructing the debate, such language reinforces the argument that surplus food distribution is the path to follow in combatting first world hunger.

Food loss and waste may be avoidable or unavoidable, intended or unintended. More specifically food waste is either deemed inedible (spoiled or rotten) or edible but unsellable (e.g., past the 'best before' date) in which case it ends up uneaten in the dump.

Surplus food refers to the redistribution of food waste potentially heading for the dumpster or landfill which on its somewhat hazardous journey from farm to fork along the food chain loses its market value and profitability and is diverted to food banks to feed hungry people. It becomes food aid protected by Good Samaritan legislation and is eaten by people.

Food aid also includes government agricultural commodity stocks available for distribution through such programmes as the Emergency Food Assistance Program administered in America by the USDA or the Fund for European Aid to the most Deprived People in the EU. These publicly financed programmes subsidize farm incomes by building national food reserves which can be used either for bolstering domestic food safety nets, for example through distribution to food banks or as overseas food aid.

Defining food loss and waste (FLW) and surplus food

More formal definitions of FLW and surplus food (Box 5.1) reveal their nuanced meanings and what connects them:

BOX 5.1 CONNECTING FOOD LOSS AND WASTE WITH SURPLUS FOOD

Food loss

'The decrease in quantity or quality of food – the agricultural or fisheries products intended for human consumption that are ultimately not eaten by people or that have incurred a reduction in quality reflected in their nutritional value, economic value or food safety' (FAO, 2015b)

'Food that spills, spoils, incurs and abnormal reduction in quality such as bruising or wilting, or otherwise gets lost before it reaches the consumer' (Lipinski et al, 2013)

'Food loss typically occurs at the production, storage, processing and distribution stages of the food value chain, and is the unintended result of agricultural processes or technical limitations in storage, infrastructure, packaging and/or marketing' (ibid)

Food waste

'the discarding or alternative (non-food) use of food that was fit for human consumption – by choice or after the food has been left to spoil or expires as a result of negligence' (FAO, 2015b)

> 'Food waste typically, but not exclusively, occurs at the retail and consump-
> tion stages in the food chain value chain and is the result of negligence or a
> conscious decision to throw food away' (Lipinski et al, 2013)
>
> **Surplus food**
>
> 'The edible food that is produced, manufactured, retailed or served but for
> various reasons is not sold to or consumed by the intended customer' (Sert et
> al, 2015)
> 'Edible, saleable food within its "use-by" date that is not sold' (Caraher and
> Furey, 2017)
> 'Food that is redirected to food banks and subsequently eaten by people' ....
> **but**
> 'Surplus food that is thrown away is considered waste. If that surplus food is
> donated or redistributed to people via food banks or other means, the food has
> avoided becoming waste. **This is why food distribution is considered a solu-
> tion to food waste**' (Lipinski et al, 2013)

Food loss and food waste are different but interrelated concepts. In their working paper 'Reducing Food Loss and Waste' for the World Resources Institute
(2013) Lipinski and colleagues describe the process by which food is lost along the
food value chain before it reaches the consumer. Food loss may well be unavoidable. Indeed, as Jonathan Bloom points out in *American Wasteland*, exposing how
America throws away nearly half of its food, food loss may result from harsh
weather, disease or insects as well as 'storage loss, spoilage and mechanical
malfunctions' and from 'inedible discards such as peels, scraps, pits and bones'
(Bloom, 2010, xii). Lipinski and colleagues would likely agree.

Food waste occurs, however, as Bloom argues, 'when an edible item goes
"unconsumed" as a result of human action or inaction. There is a culpability in
waste (2010, p.xii). It is an intentional act, either from 'negligence or a conscious
decision to throw food away' (Lipinski et al, 2013) as in the latter case when it is
sent to the land fill. Wasted food is uneaten food.

Surplus food is the result of government subsidized agricultural commodities
(surplus food stocks); of overproduction or from the overstocking of supermarket
shelves. It is edible food either for purchase or used as food aid. It can be sent
overseas or in the rich world directed to food banks, social supermarkets, school
meal programmes or charities.

Of course even food banks and their recipients may throw surplus food away. If
so it reverts to waste but on being consumed, it escapes the trip to the landfill. It is
rescued or salvaged. Surplus food is not wasted, it is eaten by people living precarious lives outside or on the edge of the market. In this way 'food distribution is
considered a solution to food waste' (Lipinski et al, 2013). As the rise of food bank
nations shows, recycling surplus food by corporate food charity is regarded as an
effective response the prevalence of food waste and food poverty in the OECD.

Prevalence of rich world food waste

As with food insecurity and domestic hunger, the prevalence of FLW also constitutes a global crisis of significant dimensions. It is estimated that 'the food sector accounts for around 30 per cent of the world's total energy consumption and accounts for around 22 per cent of total Greenhouse Gas emissions' (SDGs, 2015b, Goal 12). Yet as the UN itself has reported 'accurate estimations of the magnitude of losses and waste are lacking, particularly in developing countries. Nevertheless, there is no doubt that food loss and waste remain unacceptably high' (FAO, 2015).

A recent OECD Agriculture Policy Note agrees. While recognizing the importance of food loss and waste in the entire food chain and the need to address the issue, it reminds us that 'to date there are no commonly agreed definitions of "food waste", "food wastage" or "food loss"'. Given this lack of 'harmonization on definitions or on methodologies for food waste data collection and measurement, significant problems arise for comparative data collection and the development of targeted measures to address waste' (APN, 2016). As has been noted there are multiple visions and values of excess food (Mourad, 2015).

The OECD does not oppose the redistribution of food to feed hungry people, indeed it supports the 'minimization of waste sent to landfill, the source of significant volumes of methane' (APN, 2016). Rather the problem is the nature and measurement of FLW that has yet to be clearly defined and agreed.

However these questions are now receiving attention. A new international framework – The Food Loss and Waste Accounting and Reporting Standard – was launched at the Global Green Growth Forum (3GF) 2016 Summit in Copenhagen. As the press release reads 'The *FLW Standard* is the first ever set of global definitions and reporting requirements for companies, countries and others to consistently and credibly measure, report on and mange food loss and waste' (FLW, 2016). This has occurred as more governments, companies and other organizations are making commitments to reduce food loss and waste.

Key global partners include the World Resources Institute, the World Business Council for Sustainable Development, The Consumer Goods Forum, WRAP, EU-FUSIONS, the United Nations Environment Programme and FAO's Save Food Global Initiative as well as Big Food supporters Nestlé and Tesco. Yet despite this blue ribbon line-up advancing such a bold agenda a red flag is waving. Food waste is a symptom not the underlying structural problem which needs addressing.

Global and national food waste estimates

Bearing the FAO and OECD cautions in mind, the UN estimates a third of all food produced for human consumption in the world is lost or wasted on an annual basis – approximately '1.3 billion tonnes worth around $1 trillion ends up rotting in the bins of consumers and retailers, or spoiling due to poor transportation and harvesting practices' (SDGs, 2015b, Goal 12). Even if just a quarter of the food currently lost or wasted could be saved, it would be enough to feed 870 million hungry people in the world (FAO, 2014; Gonzáles Vaqué, 2015; HLPE, 2014; SDG, 2015b).

In high income states the cost of food loss and waste amounts to roughly US $680 billion, more than double the value lost in developing countries, despite roughly the same quantities of food being wasted in both. The FAO also finds that the rich world wastes as much food in a year (222m tonnes) as the entire net food production of sub-Saharan Africa; and that the per capita waste by consumers in Europe and North America is between 95–115 kg a year compared to 6–11 kg a year in sub-Saharan African and south and south-east Asia. Most food waste in high income countries occurs at the distribution and consumer level in middle and high income countries (HLPE, 2014, p.26).

In thinking about the prevalence of food waste and its availability as surplus food for collection by food banks, Belinda Li of the Food Systems Lab at the University of Toronto advises it is preferable to think of Edible Food Loss and Waste in relation to 'total' food available rather than total food production. 'Total' food available (Table 5.1) is more representative since it excludes feed/fuel/exports and includes imports rather than Total Food Production which includes exports and feed and fuel which can vary from 20% to 50% (BL, email communication 4.5.17).

As one can see compared to the total food available the amounts and proportion of edible food waste in high income OECD countries are considerable if not staggering. They confirm the UN estimate of one third of total food production for human consumption being lost. New Zealand's percentage of food wasted is

TABLE 5.1 Estimated edible FLW as percentage of 'total' food available, selected OECD countries and EU, Food Balance Sheets, 2007

	Total food[i]	'Total' food available[ii]	Edible FLW[iii]	FLW/TFA
	million tonnes			%
Canada	96	35	11.5	33.0%
USA	819	344	112.0	32.6%
Mexico	97	77	24	31.2%
Australia	47	16	5.9	36.8%
New Zealand	21	3	2	66.6%
Chile	23	10	5	50%
UK	49	49	13	26.5%
Turkey	91	58	23	39.6%
EU	710	395	136	34.4%

Source: FAO Food Balance Sheets (2016) http://www.fao.org/faostat/en/#data/FBS Gustavsson et al (2013); Belinda Li, Food Systems Lab, University of Toronto.
i   Includes exports/products used for feed or fuel (varies between 20% and 50%)
ii   Excludes animal feed/fuel/exports, includes imports
iii EFLW: includes agricultural post production/harvest; processing; distribution (retail/food service) and consumer waste destined for human consumption, excluding wasted inedible parts of food

much higher given it exports most of the food it produces and similarly with Chile where most of the food grown and exported are fruits and vegetables (BL email communication 2017).

These figures present a troubling picture of the scale of food waste in the rich world. It is also necessary to bear in mind that the Gustavsson et al analysis of 'Global Food Losses and Food Waste' conducted in 2011 'used the 2007 food balance sheets so this maintains some consistency with the FAO reported numbers' (BL, email communication, 2017).

National food waste snapshots

Bearing this in mind and the real difficulty of accessing reliable and comparable international data within the OECD, it is worth considering different national reports and findings regarding the food waste issue in the rich world. They vary by time periods, in numbers and interpretation but they all underline the serious nature of the issue.

USA

In the USA, it has been estimated that up to one half of all food produced is wasted (Stuart, 2009; Bloom, 2010), a figure referenced by Michael Carolan, sociologist and author of *The Real Cost of Cheap Food*, while noting the costs to the American economy of $100 billion annually (see Jones, 2005). Furthermore, Carolan writes 'US per capita food waste has increased by 50% since 1974 to more than 1400 kilocalories per person per day, or 150 trillion kilo calories per year (see Hall et al, 2009)' (Carolan, 2011).

Looking at today there is no reason to believe these figures have declined. In 2016 Suzanne Goldenberg, an environment correspondent for *The Guardian* reported new research suggesting that 'half of all US food produce is thrown away', a key reason being US consumer demand for blemish free food meaning 'much is discarded, damaging the climate and leaving people hungry' (Goldenberg, 2016a). That year, Feeding America observed 'more than 70 billion pounds of food was wasted annually, between 25–40% of food grown, processed and transported in the US' (FA, 2016; ReFED, 2016).

Canada

Meanwhile in Canada food waste is close to 40% of all food produced, of which half is said to be from consumers (Soma et al, 2016). In 2014, Value Chain Management International, based in Ontario, estimated the cost of wasted food to Canadians at $31 billion. Of note, this is a measure of the value of the wasted food itself and only represents 'approximately 30 percent of what the Canadian agriculture and agri-food system generated in 2012'. More significantly, however, VCMI notes that if the FAO costing of food waste (FAO, 2014) was to be applied in Canada by taking account of quantifiable environmental and social impacts the true scaled up cost of food waste would be $107 billion (VCMI, 2014).

EU

A 2010 preparatory (but not comparative) EC study of food waste in the EU-27, estimated total food waste generation at 89 million tonnes. It relied on 2006 Eurostat data and different national sources. The UK (14.3 million tonnes) followed by Germany, Italy, the Netherlands, France, Poland and Spain reported the highest levels of food waste in the EU-27 (EC, 2010, p.12).

A more recent 2016 study of EU food waste reported nearly 88 million tonnes of food (food and inedible parts) being discarded in the EU-28, or 20% of all food produced. This equated to 173 kilograms of food waste per person. The household (47%) and processing (17%) food sectors were deemed mainly responsible. The financial value of EU food waste has been estimated at 143 billion euros (see FUSIONS, 2016, pp.4–5). However the study's authors sound a cautionary note that 'the data behind these figures comes from different sources, which use a variety of definitions for what is considered "food waste". In addition, different studies use different methods, which can affect the resultant estimates' (FUSIONS, 2016, p.3). It has also been estimated that food wasted in Europe could feed 200 million people (FAO, 2014).

UK

In the UK, household food waste has been estimated at 7 million tonnes annually, about one third of the food purchased by consumers (Carolan, 2011). A more reccent estimate between 2013 and 2016 by WRAP (Waste & Resources Action Programme), the UK-based charity founded in 2000, puts annual post-farm gate food waste at 10 million tonnes, discarded by households and by the food industry sector as a whole – food manufacturers, retail, wholesale, hospitality and services, and households – of which 60% could avoided (WRAP, 2017a).

Certainly 10 million tonnes merits attention, particularly the estimate that household food waste (including waste to sewer) is responsible for 71% of food which is thrown away uneaten. In other words individual consumers are principally to blame. As WRAP makes clear, this is not the sum total. Excluded from this estimate is food loss and waste from pre-farm gate food production, obviously increasing the total amount of discarded food. It is omitted due to uncertainty regarding the monitoring of food waste in primary production, a newly developing area of research for WRAP and the EU.

OECD

Elsewhere in the OECD world food waste is attracting increasing attention. In Australia, food waste is recognized by the Department of the Environment and Energy as a serious problem estimated in 2013 to cost Australian households more than $8 billion each year and carrying significant environmental costs including greenhouse gas emissions (DEE, 2017). One estimate is that '7.5m tonnes of food is discarded annually in the municipal, commercial and industrial waste streams – every year' – with householders throwing away food valued at 'between $8–10bn of food a year' (Kane, 2016). FOODWISE, a national green organization campaigning to promote sustainable food reports Australians annually discarding 20% of the food they purchase: 20–40% of fruit and vegetable are rejected before they reach the shops. Indeed total food waste would fill 450,000 garbage trucks more than three times bridging the gap between Australia and New Zealand (FW, 2017).

Across the Tasman, the national volume of food wasted in New Zealand is currently being researched. WRAP, in partnership with the national waste industry organization WasteMINZ, has launched the 'Love Food Hate Waste' campaign including research into household food waste. A 2014–15 study of 1402 households across 12 different councils in the country found that 122,547 tonnes of edible food was thrown away every year with a value of $872m (WNC-NZFWAR, 2015). The campaign has resulted in a joint collaboration between 59 Councils across New Zealand to reduce the amount of food waste heading for landfill (Scott, 2016).

In Mexico, the Ministry of Environment and Natural Resources reported in 2012 that 40 million tons of garbage was being sent to the landfill each year (Godoy, 2012). Four years later, the Ministry of Social Development announced that 19 million tonnes of food was being wasted enough to feed 27 million people. It noted that Mexico's food bank network had 'offices in 29 states which recover 120,000 tons of food a year, 60% of the total in Latin America and the Caribbean' (René, 2016). In the same year it was also reported that 37% of all food produced is being lost, amounting to 10,341,000 tonnes enough 'to feed seven million Mexicans and reduce the contribution to solid urban waste in landfills that pollute soil and water' (CaribFlame, 2016).

Despite legitimate concerns about the difficulties facing the collection and analysis of food waste data and its comparability across OECD member states, the findings of global and national studies and reports raise awareness by governments, corporations and civil society about the widespread prevalence of FLW and its huge environmental, health and economic costs. Furthermore when considering the role food waste plays in driving surplus food redistribution and the stocking of food bank shelves it is helpful to explore FLW along the food value supply chain while paying attention to the upstream and downstream roles played respectively by the corporate food sector and by consumers and their supermarket shopping.

Waste along the food value chain

The food value chain comprises the various stages along which food passes on its journey from farm to fork:

- *production > handling, storage, transportation > processing and packaging > distribution and the retail market > consumption* (see Lipinski et al, 2013)

Specifically from a business viewpoint it includes 'the network of stakeholders involved in growing, processing and selling the food to consumers to eat' (see Deloitte, 2013).

- *the producers, processors; distributors, consumers* and *government, NGOs and regulators* (see Deloitte, 2013).

These are the farmers and fishers, Big Food conglomerates, food packaging and transportation companies, retailers such as supermarkets, grocery stores, farmers' markets, caterers and restaurants which make a living from food and those who purchase it. They are all concerned with food's profitability as it passes each stage towards the food trolley cart and the consumer with cash in hand. Food waste occurs at any stage when discarded food remains uneaten. The stakeholders also include the State, publicly accountable for the collective oversight of the food system and civil society as national watch dogs.

Digging more deeply from an environmental standpoint Suzanne Goldenberg (2016a,b) in tracing the lifecycle of six popular food stuffs travelling from food to fork in the USA describes six stages of food waste in explaining why so much is thrown away:

- *at the farm (potatoes)* > *before shipment (strawberries)* > *in the truck (chickens)* > *supermarkets and restaurants (leafy greens)* > *in the fridge (yoghurt)* > *at the table (bread)* (Goldenberg, 2016b).

Goldenberg's description brings to mind a series of supermarket check-out counters with each cash register indicating the upstream and downstream causes of food loss and waste.

Upstream

At the farm: Before transportation from the field begins, potatoes, emblematic of other crops, are already being discarded: too big, too little, too blemished – the cost of *failing to meet aesthetic* or *cosmetic requirements* and of *disregarding the standardization of food products* (author italics) demanded by prospective supermarkets and buyers down the line (ibid; Carolan, 2011; HLPE, 2014), in other words *growing waste* to *satisfy the food industry's need to over-produce* in order to *sell perfection* (see Bloom, 2010).

Before shipment: Strawberries, packed and ready for transportation may be already bruised due to *unpredictable weather, pest and damage from disease*.

In the truck: as Goldenberg also notes the typical food item in the USA travels nearly 1500 miles (2,400 km) from fork to mouth. On the road to food processing plants and supermarkets delays and *over-heating* may cause refrigeration to break down with truckloads of perishable or frozen foods having to be dumped. Mechanical issues with storage may also occur (Bloom, 2010).

Supermarkets and restaurants: Over-supplying by food producers and food processors, *over-stocking* and *over-ordering* by supermarkets and restaurants may well lead to perfectly good food such as bagged salads and leafy greens (Goldenberg, 2016b), as well as many other food stuffs, headed for the landfill unless they find their way to food banks. *Market concentration in the supermarket sector* means food processors and farmers who are con-tractually tied to one supermarket and unable to sell elsewhere may need to recycle rejected food back into the ground (Stuart, 2009) with *over-production being encouraged* when 'forecast-orders' exceed what is bought on the day (Carolan, 2011, p.130).

Downstream

In the fridge: Consumer over-spending, promoted by advertising and two for one deals (BOGOF), results in over-stocked fridges and freezers and the problem of uneaten 'left-overs' and forgotten frozen items being thrown away due to *the confusing array of date labels* particularly for such products as milk, yoghurt and other dairy products (Goldenberg, 2016b). *Over-spending* is also encouraged by the relative *cheapness of food* (Carolan, 2011; HLPE, 2014) though ironically even 'cheap' food is unaf-fordable for millions of people. *Overconsumption* of food is also detrimental to our health and the environment and 2 billion people globally are overweight or obese (SDGs, 2015b, Goal 12).

At the table: What then to do about all the *'left-overs'* and *uneaten bread*, 'typically one of the most wasted items at restaurants … with families often *buying more* than they can use' (Goldenberg, 2016b). No wonder that 'many food banks have too much of certain foods such as bread and pastries, particularly in the USA' (Mourad, 2015, p.75). Over baking and overproduction is clearly evident in the UK which, it has been reported, 'wastes an estimated 44 percent of the bread it produces, adding up to 24 million slices per day' (Martinko, 2017). There is only so much bread which food banks can distribute.

Goldenberg's analysis of the stages of food loss and waste up and down the food value chain highlight their complex and interrelated causation. Yes there is una-voidable food loss when crops perish in the field but the reasons for food waste are principally man-made, those of *homo economicus:* corporate *over-production, growing waste to sell perfection; over-baking; over-supplying; over-ordering; over-stocking backed by market concentration in the supermarket sector*. Clearly there is more than enough for food charity and surplus food redistribution.

Blaming the consumer

At the same time the finger points at consumer behaviour and household food waste: *over-spending, over-buying; over-stocking* fridges and freezers; and *over-consumption*. As the 2030 SDG Agenda notes 'while substantial environmental impacts from food occur in the production phase (agriculture, food processing)' …. 'households influence these impacts through their dietary choices and habits. This consequently affects the environment through food-related energy consumption

TABLE 5.2 Estimated consumer food waste as percentage of edible food waste,[i] selected OECD countries and EU, FAO Food Balance Sheets, 2007

	Edible food waste	Consumer food waste	CFW/EFW
	million tonnes		%
Canada	8.2	5.0	60%
USA	82.0	52.0	63%
Mexico	16.4	3.7	22.5%
Australia	2.3	2.2	57%
New Zealand	0.8	0.4	50%
Chile	2.9	0.5	7%
UK	10.1	5.8	57%
Turkey	14.8	2.9	19.5%
EU	88.0	47.0	53%

Source: FAO Food Balance Sheets (2016) http://www.fao.org/faostat/en/#data/FBSDerived from Gustavsson et al (2013); Belinda Li, Food Systems Lab, University of Toronto

i   Food destined for human consumption but excluding agricultural post-production food loss and wasted inedible parts of food

and waste generation' (SDG, 2015a), in other words as consumers we bear a primary responsibility for food waste. As Bloom writes 'home is where the waste is' (Bloom, 2010). Yet how fair is this?

As Table 5.2 indicates consumer food waste in the OECD food bank nations as a percentage of edible food waste is no small matter, in fact the data is shocking with North America leading the way. If there is any consolation it is that the percentages for Mexico, Chile and Turkey are appreciably lower, perhaps due to these less affluent countries having more regard for the value of the food they grow and eat.

That being said there is little wonder that the FAO attributes the causes of food waste in wealthy nation states primarily to consumer behaviour whilst noting that public policies and regulations designed to address priorities in other sectors including subsidized agricultural commodity stocks – to support farm incomes or food assistance programmes – are also part of the problem. They are intended to produce surplus food some of which may become lost or wasted. An unintended consequence of 'of food safety and quality standards may also remove perfectly safe food for eating from the food supply chain'. In blaming consumers, the FAO notes 'inadequate planning of purchases and failure to use food before its expiry date also leads to avoidable food waste' (see FAO, 2015b).

Still, in the face of the food industry's relentless advertising how fair is it to hold consumers responsible for scandalous food waste? What are they individually expected to do in taking on a dysfunctional waste producing industrial food system described by Tim Lang as the 'structured mismatch between production, consumption, environment, health and social values' (Lang, 2013).

Despite being blamed for food waste, one response which consumers may take is to become food bank volunteers. Ironically, however, in handing out surplus food they unintentionally function as a safety valve ensuring that corporate food waste carries on. The irony is further compounded as recovering downstream consumer food waste presents a significant organizational challenge to food banks. Furthermore while hunger may be alleviated for some the volunteers free labour ensures corporately captured food charity remains entrenched in a food industry known for its low wages (Scott-Thomas, 2013).

In light of Big Food's upstream dominance of the food system it is well to ask how corporate food waste shapes the argument about surplus food redistribution.

Food waste manufacturing the surplus food agenda

Since the origins of St Mary's Food Bank in 1967, the challenge of preventing unused edible food ending up as food waste in the local dump has informed surplus food distribution in the rich world. Food industry support and corporate food waste certainly was instrumental in the rise of Feeding America which today is focused on keeping rotting greenhouse gas producing food waste out of landfills noting on its website that 'more food reaches landfills and incinerators than any other single material in municipal solid waste' (FA, 2017f). Food waste manufactures and drives the surplus food agenda

Environmental protection

Significantly the food waste-to-surplus food argument has the backing of the US Government through the influential voice of the Environmental Protection Agency (EPA). The EPA claims that 'by keeping wholesome and nutritious food in our communities and out of our landfills, we can help address the 42 million Americans that live in food insecure household' (EPA, 2017). Second to reducing food waste at source, the first priority on its Food Recovery Hierarchy, is donating surplus food to food banks for feeding hungry people. Other recommendations include converting food waste into food scraps into animal feed; for industrial uses; composting and landfill/incineration (EPA, 2017)

Stocking food bank shelves with surplus food is a powerful environmental message emanating from the leading federal authority on this matter. Despite the Trump Administration's climate change scepticism and rejection of the Paris Accord, the EPA Food Recovery Hierarchy will undoubtedly remain a central and influential set of national guidelines.

Big Food's greening food waste message

Within the US and Europe the environmentally driven food waste argument promoting surplus food as a response to rich world hunger has gathered food industry momentum. Feeding America has strengthened its commitment to corporate food waste reduction. It is now a member along with Waste Management, Inc of the national Food Waste Reduction Alliance (FWRA) jointly established in 2011 by the Grocery's Manufacturers Alliance, the Food Marketing Institute and the National Restaurant Association to reduce food waste, increase safe and nutritious food being donated to those in need and recycle unavoidable food waste from landfills. The FWRA is backed by Big Food companies such as Campbells, The Cheesecake Factory and ConAgra (FWRA, 2017; FA, 2017f).

As noted previously the European Federation of Food Banks with EU Food-Drink Europe and EuroCommerce, both international retailing associations, jointly prioritize food waste reduction. The UN SDG 12.3 goal of halving global food waste by 2030 (see WRI, 2017) is shared by FEBA (2017e), the Global Food-banking Network and the H-E-B Food Bank Leadership Institute.

Corporate social responsibility and zero waste

Inevitably Big Food and its supermarket chains have corporate and financial interests to consider when responding to unacceptable levels of global and national food waste. However WRAP's 'Love Food, Hate Waste' campaign in the UK in partnership with the British Retail Council (BRC) is actively driving the corporate social responsibility argument that 'zero waste' and surplus food distribution are not only good for business but essential for stocking food bank shelves (WRAP, 2017b; BRC, 2016).

Indeed, WRAP in promoting the Courtauld Commitment 2025, launched in 2005 and directed at a 'zero waste economy', has successfully engaged Big Food in its implementation. Cadbury, Mars, Nestlé, Unilever and BRC-leading supermarkets such as Tesco, Sainsbury's, Asda, Morrison's, Marks & Spencer, Waitrose and The Co-op have committed to the food waste reduction cause (HLPE, 2014; BRC, 2016). Not only did they commit to 'drive down food and drink waste by a fifth within the next decade' (Wood, 2016), but also, in the interest of transparency to provide data regarding food loss and waste in the overall food supply chain (HLPE, 2014). Indeed in 2013, ASDA (the UK supermarket chain owned by US giant Walmart) had already unveiled a multi-million pound pledge to redistribute all surplus stock to directly to FareShare (Smithers, 2013). The Love Food Hate Waste has also been influential in Australia, New Zealand and the USA.

Ahold Delhaize (AD), the Netherland's based international food retail giant, may lead the field in corporate social responsibility. Since 2011 it has provided data on its food loss and waste as 1–2% of its total food sales (HLPE, 2014). Yet, its on-line *Responsible Retailing Report 2015* goes much further in terms of transparency and accountability in reflecting on bigger picture issues which it faces as a major player in the corporate food sector.

AD's 2016–2020 report card indicates its progress towards offering more healthy and affordable foods; addressing food security in part through food donations; and flagging global warming by caring for the environment by reducing its carbon foot print (AD, 2015). It is a prime example of the food waste-surplus food-poverty reduction strategy approach of the food retail sector partnering with food banks.

Food waste arguments are driving surplus food redistribution onto political agendas as evident in other European countries. Legislation introduced in France made it the first country to ban food waste in supermarkets compelling them to donate to food banks or charities. Italy has introduced food waste legislation and Denmark has played a leading role.

Social supermarkets (community food shops in the UK), described as a 'win-win-win' for Europe's poor have also expanding rapidly. One report indicates there are 'about a 1000 such stores spread across the continent, including France, Austria, Belgium, Luxembourg, Romania and Switzerland' (Graslie, 2013). Even in the Nordic welfare states, a pilot study for the Nordic Council of Ministers is recommending that that 'foodbanks classified as food business operators in a legal context' have an underutilized potential to reduce food waste (see Norden, 2015a, b).

Caution and a red flag

Nevertheless caution, even a red flag is in order before assuming the 'win-win' effectiveness of surplus food redistribution in reducing food waste, let alone domestic hunger. Table 5.3 shows what a tiny proportion of edible food waste is likely to be recovered by food banks. Admittedly to ensure consistency the food waste figures are for 2007 and the years of food bank recovery data for availability reasons vary. No doubt recovery methods have improved over the last decade but the data shows what an immense challenge faces surplus food recovery if it is to have a significant impact on reducing food waste.

Of course, there being an endless supply of food waste it can be argued that this is a challenge which corporate food banking should accept. Yet what appreciable impact will ramping up surplus food redistribution have on resolving first world hunger. After all corporate food charity depends on voluntary labour and how well placed and committed are food banks with their anti-hunger focus to solving the epidemic of food waste in the rich 'throwaway' world in which we live.

TABLE 5.3 Estimated recovery of EFLW (post-harvest, processing, distribution) by food bank associations, Canada, USA and Europe, 2007–2012

	EFLW waste	Amount recovered[i]	% recovered
	metric tonnes		
Canada	3.2m	3,306	0.1%
USA	30m	962,000	3.2%
Europe	41m	388,000	0.94%

Source: FAO Food Balance Sheets (2016) http://www.fao.org/faostat/en/#data/ FBS Gustavsson et al (2013); Belinda Li, Food Systems Lab, Vancouver

i   Annual Reports: Food Banks Canada (2008); Feeding America (2007); European Federation of Food Banks (2012)

The 'throwaway society'

Over-production; over-ordering; over-stocking; over-supplying; over-purchasing and over-eating are all products of the abundant throwaway rich world, described in the USA as living in a culture of waste (Bloom, 2010). This is deeply rooted. As Tim Lang has noted 'modern societies have a problem with waste. The entire economy is wasteful, a distortion of needs and wants' (Lang, 2013), a culture of wasteful consumption served by over-production and over supply and advanced at every commercial opportunity by relentless advertising. A recipe for the easy life.

As a respondent to a survey in Toronto commented when asked why people are wasting food, 'I think a lot of it is probably generated due to (the) convenience economy. We have to have everything available at all times so stuff sitting there at the supermarket that doesn't sell in a couple of days has to be thrown out. I think that is driving a lot of food waste' (Soma et al, 2016). In this context food waste is just par for the course given that built-in obsolescence is hard-wired into our systems of production and patterns of consumption always seeking the latest gadget or technological device which will fix problems and make life easier and happier. The question is whether food waste is a quick fix for food poverty.

Corporate food waste, symptom not a cure

Food waste keeps piling up because it is a symptom, and not *the* problem which structurally needs to be addressed: the dysfunctional and unsustainable global agro-industrial food system. In a recent article, Felicity Lawrence points to the huge environmental/ecological damage to global food production caused through the over-use of agrochemicals and GM seeds by the top six giants in this field – Bayer, Monsanto, Dupont, Dow, Syngenta and BASF. She concludes that this agro-industrial way of producing our food is broken. Its ecological and health consequences have been steep losses in biodiversity and of pollinators vital to food; and a failure to ensure the long-term provision of healthy and nourishing food while 750 million people are still hungry and nearly two million are overweight or obese (Lawrence, 2016).

Yet, even greater market concentration is ruling the day. The six giants are undergoing mergers becoming 'even bigger and more powerful' players in controlling world food production, supplies and prices. While these companies talk of the threat of climate change to food security, 'the agroindustrial food system is one of the most significant causes of it, contributing a third of all manmade greenhouse gas omissions' (Lawrence, 2016). Given the enormous challenges of what to do about the upstream problems of the failed food system it is necessary to point out that the food waste–surplus food redistribution argument is promoted by the six agro-business giants as significant players in Foodbanks Inc either partnering with or donating to corporately captured food charity.

The *downstream* matter of the annual discharge of millions of tonnes of corporate food waste into our 'food secure' wealthy societies remains. It certainly demands an answer but that will not happen unless or until the *upstream* matter of the failed and disconnected agro-industrial food system is fully recognized and transformed. Meanwhile food banks in unholy alliance with their corporate partners (see Fisher, 2017) look on and governments look the other way.

Reflections

Certainly in times of austerity there will be strong public support for feeding hungry people. The moral imperative is clear particularly when buttressed by the reality of mountains of surplus food otherwise destined for landfill. In that context taking action to advance an ethical and sustainable environment and protect life on planet earth goes without saying.

However claiming that surplus food – as the product of manufactured corporate food waste – is the answer both to domestic hunger and food waste itself is a mistaken prognosis. Food waste and surplus food are symptoms of a broken global industrial food system. Domestic hunger and food banks are symptoms of broken social safety nets and dismantled welfare states. Who stands to benefit from such remedial treatment: those standing in the breadlines, the food industry, food con-sumers, tax payers, civil society, government or citizens. Both are complex structural issues requiring more than doses of recycled food waste and corporate charity.

We need instead to pose the counter-intuitive question whether the rise of food bank nations and corporately sponsored food charity is part of the problem and not a solution to domestic hunger. How effective are they? Do they act in solidarity with the poor? Are they an obstacle to a food secure nation?

6

CORPORATE FOOD BANKING

Solution or problem

The social construction of hunger as a matter for charity allows us to believe there is very little that governments can do, assuming they wish to, in addressing the problem of widespread food insecurity prompting the question as to whether food charity in the rich world is part of the solution or part of the problem and, indeed, where accountability does and should lie. The rise of food bank nations and the corporate capture of charitable food banking and safety nets strongly signals the de-politicization of domestic hunger and food insecurity as issues demanding the priority attention of nation states. The retreat to Victorian age charity seems well entrenched.

Thinking about food charity as the primary response to hunger

In early 2007, with spring barely arrived, this much was confirmed for me one bitterly cold and rainy morning. Judy Graves, Vancouver City Council's full-time Advocate for the Homeless, had invited me to walk the streets in the Downtown Eastside (Canada's poorest postal code) to meet those standing in the bread lines and sample the surplus or donated charity food on offer at a number of its missions, soup kitchens and meal programmes.

Regularly treading this path was part of Judy's watching brief in this urban neighbourhood where she was well-known as a friend and advocate for all who lived their lives on the street. It was an invitation to think about the provision and experience of emergency food assistance at its last desperate point on the unjust food supply chain. It was a morning of contrasting extremes. In 2007 wealthy west coast Vancouver was selected the world's most liveable city, a ranking it was to maintain for the next five years (NP, 2011).

In the Downtown Eastside practical compassion was much in evidence but only a bare minimum of an uncoordinated food safety net and little of what might be called nutritious and healthy food on offer. We were always invited in, never refused, yet as we slipped from food line to breakfast programme the stark reality of the quality of the food on offer (more likely a cup of over-salty noodle soup or broth mixed with tortellini and frozen mixed vegetables) led us to write an article entitled 'Let Them Eat Starch' (Riches and Graves, 2007) for *The Tyee* BC-based independent online magazine. What struck me most was the anonymous on-line respondent who replied that the article's most telling statement was that 'the many acts of charitable goodwill prevent us from asking who is accountable for this morally unacceptable state of affairs. We go on giving and giving but the situation never changes' (ibid).

A highly pertinent question, yet, should we be surprised if no one, then as now, had any expectation that government with its constitutional and legislative powers would be taking action? After all, during a neoliberal era when income and wealth inequality had been growing, and to this day continues to widen we have come to accept the normalization of local, community-based emergency food aid and a legitimate role for corporate food banking as the food custodians and protectors of the poor. Not coincidently the Global Foodbanking Network was formed in 2006 and two years later Matthew Bishop and Michael Green were to publish *Philanthrocapitalism: How the Rich can Save the World* (2008).

Big philanthropy

This should not be a surprise. In this Gilded Age of neoliberalism, corporate charity, or to use the language of the global super rich, 'philanthrocapitalism' or 'venture philanthropy', holds sway. Billionaires, celebrities and transnational corporations have been persuaded to raise their level of philanthropy and charities in a *quid pro quo* to take on the best practices of their business and financial masters. All the while welfare states and publicly funded social safety nets are being dismantled and institutionalized food charity, we are told is becoming more efficient and effective. More and more hungry people in the rich world are being fed. Given this resurgence of charity and the decline of public policy in addressing social need there are however awkward questions for powerful wealthy philanthropists and corporations seeking to do good.

As Benjamin Soskis, writing in *The Atlantic* about what might be termed Big Philanthropy in modern day America, reminds us there is a need, even a democratic imperative to criticize public benefactors and philanthropy. To do so, he writes 'is not a mark of incivility', going on to say 'there is now a significant and vocal faction willing to call philanthropic ambitions into question. In part, the pushback can be traced to the nation's mounting uneasiness with income inequality and to the spread of economic populism that refuses to regard the concentration of wealth charitably' (Soskis, 2014).

Sociologist Linsey McGoey in her book *No Such Thing as a Free Gift*, an analysis of the Bill and Melinda Gates Foundation and its role in international aid, also takes aim at philanthrocapitalism. Referencing Bishop and Green's earlier study, she identifies its two key features: the importance of charities to emulate the way 'business is done in the for-profit capitalist world'; and to protect the idea that 'capitalism itself can be naturally philanthropic', driving innovation in a way 'which tends to benefit everyone, sooner or later, through new products, higher quality and lower prices' (see McGoey, 2015, p.7; Bishop and Green, 2008, p.194). In other words, in the context of fighting hunger, manage your food bank according to the business principles of leveraging and best practices and in the end enlightened corporate self interest will combine with the public interest and all the hungry will be fed. If this sounds like the credo of *homo economicus*, it is.

This is not to argue that all charitable food banks are run according to this credo. However, in light of charitable food banking's corporate capture by powerful and immensely wealthy Big Food players, it is not unreasonable to assume Big Philanthropy is looked to as the food provider of last resort in a neoliberal culture in which governments outsource and down download their social obligations and responsibilities to charity and the private sector. As Tim Lang has commented in reflecting on Britain's 'destructive addiction to food banks', why is it that 'rich societies turn to crude and inefficient charity to paper over' the cracks in the food system? (Lang, 2015). One might also ask why this belief that corporate food charity holds the answers to broken publicly funded social safety nets.

With that in mind McGoey's trenchant criticisms of philanthrocapitalism invite attention. She points to its lack of accountability and transparency; the channelling of private funds towards public services which erodes support for government spending on health and education; and the view that 'many philanthropists have made their fortunes through business strategies that greatly exacerbate the same social and economic inequalities that philanthropists purport to remedy' (McGoey, 2015 pp.8–9). Admittedly she is writing about international aid supported by the vast endowed fortune of the Bill and Melinda Gates Foundation, valued at US $44.3 billion in 2014 (BMGF), and its three trustees Bill and Melinda Gates and Warren Buffet. The Foundation provides 10% of the World Health Organization Budget, donating more than the US Government (ibid).

In the context of food poverty and income inequality in the OECD, McGoey's critique when applied to hunger as a matter best left to charity, raises similar questions. Should we be asking whether such corporate social responsibility might be better understood as corporate social investment with profitability being safeguarded while the causes of food poverty remain unaddressed? To what extent has the corporate promotion, funding and institutionalization of food banking led to the politics being taken out of hunger. Moreover who holds the charitable and corporately dependent food bank industry accountable for its claim that it is part of the solution to food insecurity?

An effective solution?

To some, however, it may seem a waste of time to question the role of charity in combatting hunger and the need for emergency food aid, especially in affluent societies. There is a strong belief that food banks are part of an effective solution, perhaps even the solution. After all poverty is with us always and altruism is deeply embedded in all cultures and religions and concern for others is basic to civilized living around the globe. The moral imperative to feed hungry people expresses a common humanity which informs practical compassion and the responsibility to offer a helping hand to strangers who are unable to feed either themselves or their families. Against a backdrop of Big Philanthropy, Big Food's corporate social responsibility and the tireless work of countless volunteers it is perhaps of little surprise that 60 million people are now said to be the beneficiaries of food banks in high income nations (Gentilini, 2013).

A safety net for the safety net

Interestingly Ugo Gentilini, a Social Protection Specialist at the United Nations World Food Programme in his paper on *The State of Food Banks in High-income Countries* (2013) writes that food banks 'complement the more formal, public safety net system provided by the state' and indeed could be considered 'a safety net of the safety net' (ibid). Acknowledging the need for a debate about roles and responsibilities of 'downstream' providers of food aid and recognizing that the 'upstream' causes of poverty, food insecurity and social exclusion require attention, he concludes that 'food banks are an instrument, not an end in themselves. As such they are part of the solution, not the problem' (ibid).

If one understands effectiveness as an immediate, practical and successful response to alleviating hunger in the face of emergency food needs this view is widely shared. It is indicative of the public support which international charities such as Caritas and the Salvation Army receive as also the Global Foodbanking Network and the European Federation of Food Banks through a myriad of national and local food banks, food pantries and meal programmes. The Food and Agriculture Organization (FAO) the World Food Programme and Big Food are key participants (H-E-B/GFN-FBLI, 2015) and certainly understand food banking as part of the solution.

Why not? As Figure 6.1 shows the language describing the motivations and work of national and global food bank organizations, selected from mission statements, websites and annual reports, sends a confident and uniform message promising effective results.

Alleviating or relieving hunger is certainly central to the food bank movement's objectives of collecting and delivering surplus food as emergency food aid. Their declared aims likewise speak of fighting to end hunger, calling for solidarity, building partnerships and solving the hunger and food waste problems. These are stirring and influential words suggesting that food poverty and food waste will be fixed and food banks only a temporary measure. They imply that food insecurity and food waste in food abundant countries are technical problems and donating or supplying surplus food will be a 'win-win' solution to both, and that food banking is central to the answer.

Associations	Food banking
Feeding America	*'feed America's hungry', 'reduce waste', 'together we can solve hunger'*
Food Banks Canada	*'relieving hunger today, preventing hunger tomorrow' 'policy advocacy'*
BAMX (Mexico)	*'fight against food waste and hunger', 'free food', 'partnerships'*
Food Bank Australia	*'fighting hunger', 'food relief'*
Trussel Trust (UK)	*'stop hunger', 'break cycle of poverty', 'stop need for food banks'*
FareShare (UK)	*'solution to hunger; 'no good food should be wasted'*
FEBA	*'reducing hunger', 'fighting food waste', 'calling for solidarity'*
GFN	*'proven solution to global hunger and food waste'*
H-E-B/FBLI	*'solve the interrelated problems of hunger and food waste'*

FIGURE 6.1 Missions of national and international food bank organizations, 2013–17
Sources: Food bank web sites, annual reports 2013–2017

What charitable food banking brings to the table

Certainly, over time, food banking in OECD member states has stayed the course, a marker one could suggest as evidence of its success. Evidence of slowly evolving food safety nets is not hard to find. Charitable food provisioning led by non-profit NGOs and innovative social entrepreneurs has adapted to the need for providing more fresh, nutritious and healthy foods; worked with collective kitchens; expanded to include food distribution hubs; seen the development of social supermarkets; expanded food share operations and good food boxes; introduced food vouchers for use at farmers' markets; developed school meal programmes and even widened access to emergency food through the use of food apps. The waste not/want not motivation has been a driving force.

Importantly, as anthropologist Pat Caplan has observed, 'the performance of charity, in the form of provision of food aid, arises from one of our most important attributes: our humanity, and a recognition, however imperfect, that this is shared by the "Other". Far from denying that the situation of food poverty exists, the giving of food aid by volunteers attempts to grapple with it, and therefore cannot be summarily dismissed as just part of the problem rather than part of the solution' (Caplan, 2016, p.9). In other words, a more nuanced understanding of the role played by food banks in addressing domestic hunger is necessary.

In this respect research fellow and food bank author Hannah Lambie-Mumford's critical study of the notion of success in emergency food provision from the perspective of an ethic of caring is significant. It sheds light on the ways by which caring as an end in itself is critical to the success of such provision, offering not only food in times of crisis but places of safety, and through the act of giving provides gateways to other types of assistance, what is termed signposting or referral to other social agencies or wider support systems (Lambie-Mumford, 2017, pp.132–135)

Advocacy and public policy

As their websites demonstrate national food bank organizations such as Feeding America, Food Banks Canada, Foodbank Australia and the UK's Trussell Trust have incorporated prevention, research and policy advocacy into their agendas, an indicator of successful organizational development.

The advocacy role is one which has been taken on by Feeding Britain, an independent charity set up in 2015 as result of the report of the All Party Parliamentary Inquiry into Hunger in the United Kingdom (2014), funded by the Archbishop of Canterbury's Charitable Trust. Acting on the APPG's model of Food Bank Plus, Feeding Britain has set itself the hard task of working towards a hunger-free Britain by 2020. It aims to do this by 'coordinating a series of projects to relieve and prevent hunger, and working for reforms at a national level to reduce the nations vulnerability to hunger' (APPG, 2015).

While at arms length from front line action, in this process Feeding Britain is partnering with food banks across the country. It is seeking at one and the same time to relieve the symptoms of hunger and to eradicate its causes. This is a formidable challenge but not dissimilar to those of the food bank industry which has long seen itself addressing immediate need and simultaneously finding the solution to long-term food insecurity.

Yet when it comes to the matter of food waste questions arise about the focus of food bank advocacy directed at public policy in terms of preventing hunger. Lindsay Boswell, CEO of FareShare makes the cogent argument that it is cheaper for the food industry to send its food waste for anaerobic digestion than it is to deliver it to food banks with the goal of feeding hungry people. He believes 'the Government can make an impact to reduce the food waste mountain by acknowledging the existence of these costs, and introducing fiscal or other financial incentives for food redistribution so the food can be saved and provided to people across the country' (Boswell, 2015). He cites France, Portugal and America where such financial programmes, presumably tax incentives, have been successfully introduced. This however will do nothing to prevent hunger and poverty, but will only, as American experience shows, further embed the corporate food charity safety net.

However, Mark Winne, former executive director of the Hartford Food System in Hartford, Connecticut, says it 'good news that food banks (in the USA) are beginning to take public policy approaches more seriously' (Winne, 2008, pp.75–77), a point echoed by Jessica Powers, formerly of WhyHunger, who recently pointed to the need for more recognition of 'an increasingly skeptical and vocal critique of the status quo' from within the charitable food sector, in the US and Canada (Powers, 2015b; see also Powers and Cohen, 2016).

As Winne further commented the influential people who sit on the boards of food banks, rarely 'promote a vital discourse around hunger, food insecurity and poverty.... Generally speaking, they do not, because influential people don't attain exalted positions within a community's hierarchy by asking hard, controversial questions or by becoming agitators. Upsetting the apple cart is not the way it is done in polite society'. Rather he argues that charitable food banking is about managing poverty (Winne, 2008, pp.75–77) which can only appeal to governments (Winne, 2008, pp.75–77).

A silent tool of government

There is little doubt that austerity minded governments approve surplus food redistribution and food charity as effective responses to a perceived hunger/food waste problem. Inexorably and with little or no public debate charitable food safety nets have been quietly institutionalized at little or no cost to governments always looking to lower taxes. Public policy is as much about the choices governments make and do not make. While governments for the most part remain silent about the food poverty question they give corporate food charity their implicit and at times open support. How else to understand the US Congress's food bank tax incentive legislation, the high priority given by the Environmental Protection Agency to redistribute wasted food to food banks or the French Government's enactment banning supermarkets from sending surplus food to the landfill. At the same time the EU's former MDP and now LEAD surplus food aid programmes have consistently supported food banking as effective agencies for food redistribution.

Senior politicians, of all political stripes, lend credence to the idea of food banking being a solution, however temporary, to domestic hunger. When Hazel Hawke, the partner of then Labour Party Prime Minster of Australia opened the Melbourne Food Bank in 1994 a message was sent that food banking was an effective approach to food insecurity. Similarly, when the President of Mexico acknowledged the participation of Mexico in the creation of the Global Food-banking Network (Castrejon Violante, 2017b). More recently, in 2016, the new Liberal Party Prime Minister of Canada Justin Trudeau and his partner Sophie Gregoire-Trudeau were seen on television handing out food parcels at the Moisson Montreal Food Bank. Food banking receives political endorsement from the very top. It is a silent no cost policy choice for the State.

The same holds true in the UK. When former Conservative Prime Minister David Cameron was in Opposition, he challenged the then Labour Administration over the case of a single parent parcel living in the fourth richest country in the world being dependent on a Salvation Army food parcel. After assuming power he changed his tune. In 2015 he allowed Job Centres to place their advisors in food banks. His Minister for Work and Pensions 'welcomed food banks' and his Minister of Employment described them 'as playing a vital role in welfare provision' (Garthwaite, 2016, pp.57–61). This was the Big Society at work. Such governmental support echoed Canadian research from thirty years earlier that food banks had become a second tier of the welfare system (Gandy and Greschner, 1989). Today they are reported to be a core part of welfare systems in Australia and Spain (Forcado et al, 2015), as surely they are in other OECD member states, all examples of social policy by neglect.

Stop gaps – successful failures

Yet, as Olivier De Schutter, the former UN Special Rapporteur has observed 'food banks can never be more than a stop-gap, and can only offer basic subsistence from day to day – and not a route out of poverty' (De Schutter, 2013). It is certainly correct food banks cannot and do not guarantee subsistence, let alone an adequate diet and healthy eating each day of the month. WhyHunger, the leading US anti-hunger organization, recently published a special report by Jessica Powers acknowledging America's food banks as saying that charity will not end hunger (Powers, 2016), to be fair a position likely shared by other national food bank organizations.

Yet, what is one to make of the view that food bank volunteers in Wales feel uneasy about the work they are doing and would like to see food banks phased out (Lang, 2015), a concern shared by many UK volunteers. Or research by Hannah Lambie-Mumford which finds 'very often volunteers at food banks will say "We wish we didn't exist: our ultimate aim is do to ourselves out of business"' (McBain, 2015). In Canada there is even a union of food bank volunteers, Freedom 90, based in Ontario. Its members all wish to retire from their charitable activities before they reach 90 years of age. They believe food banking continues to brand poverty as 'hunger' and 'mask its causes': inadequate incomes which are due to low wages, precarious work and social assistance levels are too low to provide adequate food and housing (F90, 2017). Government, the volunteers believe, is publicly responsible for ensuring people's basic needs are met.

As for food banks being a part of the solution this is a mistaken assumption. While their staying power, insititutionalization and corporative capture may seem to demonstrate effectiveness, at best they should be understood as has been argued (Ronson and Caraher, 2016) as 'highly visible successful failures'. As they suggest food banks 'offer a valued service to those with insufficient resources to make ends meet but the efficacy of their services is rarely questioned' (ibid.). Indeed who stands to benefit most from the replacement of income security and publicly funded social safety nets by charitable and fragmented food safety nets with corporate food banking at its helm? Surely food banks are part of the problem and not a solution to ending food poverty and food waste?

A continuing problem?

Corporate charitable food banking claims to alleviate domestic hunger and offer 'win-win' solutions to food insecurity and food waste in high income countries. They are presented as a practical and effective emergency response. Yet, as Olivier De Schutter has argued 'food banks should not be seen as a "normal" part of national social safety nets. They are not like cash transfers or food vouchers, to which people in need have a right under developed social security systems' (De Schutter, 2013). Nevertheless food banks have been established and are on the ground running. It is therefore important task to question their effectiveness.

Certainly their visually attractive websites and annual reports provide impressive numbers regarding the millions of people who have been fed; the huge volumes of wasted and surplus food collected, donated and distributed; the millions of meals provided; the financial support and management expertise received from corporations, foundations and ordinary folks and yes, quietly (maybe ashamedly) aided and abetted by governments in the form of food safety and food waste legislation, grants and tax exemptions. The GFN claim that food banks are the link between food waste and hunger is widely promoted. It is therefore not difficult to believe the Big Food dependent food bank movement is effectively alleviating domestic hunger and solving the problem of food insecurity and food waste in the OECD world.

After all it is rooted in the idea of neighbour helping neighbour and the principle of practical compassion, the moral imperative to feed hungry people. Food banking and its safety nets are institutionalized and globalized. The boards of its national organizations include knowledgeable and experienced members of the transnational corporate food sector. Paid staff (some with outrageous salaries) conduct its business and a massive army of committed volunteers provide free labour. Indeed from an environmental perspective food waste reduction legislation can only assist food banking being perceived as an effective and efficient anti-hunger food redistribution safety net (Macdiarmid et al, 2016). Yet how solid is food banking's claim of effectiveness.

Alleviation for some but inability to meet demand

The fact of the matter is that the scale of the food insecurity and domestic hunger problem in the rich world far exceeds the capacity of charitable food banking dependent upon wasted and surplus food and the embrace of the corporate food sector. As has been noted in the USA, the birthplace of modern food banking; it is the public food safety net which provides the significant majority of the nations food assistance (Poppendieck, 2014a), with only one in 20 bags of food assistance coming from charity which spends $5.2 billion compared to $102 billion by the Federal State (see ch, 2, BW, 2014), even then the prevalence of food insecurity remains remarkably high. In Canada of those officially counted as food insecure only one in four use food banks and of those who do many still go hungry, whilst in Europe, the figure stands approximately at one in eight (Ch 2, Table 2.3).

It is necessary to bear in mind that the food bank usage data are underestimates of the real numbers of people experiencing food insecurity and that official estimates, with the exception of Canada and the USA, are not routinely and robustly collected. In the OECD the data is missing. When they are, as in the EU, the measures of food insecurity do not capture the full experience and widespread prevalence of people being unable to feed themselves and their families.

Nevertheless, despite these omissions and shortcomings it is reasonable to conclude that for some people, albeit a significant minority of people who are going hungry, food banks do provide a degree of alleviation while many more millions are excluded. Put bluntly, food banks, however well organized, are unable to meet demand.

Running out of surplus food

As I am writing this today in the small Vancouver Island town where I live there is a photo in the local newspaper of the Salvation Army food bank showing empty shelves with an appeal for food donations. This is a historical, troubling and continuing problem for food banks large and small whether or not directly dependent on Big Food or local grocery stores and restaurants for their supplies of surplus food. It is the reason why food drives and fund raising remain integral to the whole food charity operation. The photo and appeal is emblematic of the problem of food banks everywhere, however efficiently organized, always running out of food.

Another example from Germany is one of food banks and food panties struggling to meet demand. The reasons according to a 2014 article in *Der Spiegel Online* are that people are less inhibited about asking for free meals and relying on charity. More students are using food banks plus the concern that free movement will be attracting low wage workers. Also it is argued food banks have expanded too quickly and turned away from their original objective of simply feeding the needy. There has been a decline in provisions, increased competition amongst food banks seeking food and the introduction of charging those using food pantries small amounts as a way of raising funds to purchase food (Kleinhubbert, 2014).

At the same time there is also limited 'take-up' of food bank aid. As Elaine Power has written 'the reality is that the majority of hungry Canadians never even get to a food bank' with less than one in four of Canada's food insecure resorting to this form of food charity. She goes on to say 'even worse, receiving a food bank hamper does not change one's objectively measured food security status. In other words, food banks do not create food security for their clients. That's because food banks run on donations and are only able to provide what is donated, sometimes supplemented with purchased food' (Power, 2011). Whilst empirical evidence is lacking in other food bank nations, it is likely this predicament is widely experienced

Nutritional inadequacy – public health emergency

On both sides of the Atlantic and in Australia concerns have been expressed from a nutrition and public health perspective pointing to the ineffectiveness of food banks as a response to food insecurity. The Ontario Society of Nutrition Professionals in Public Health (OSNPPH) in a Position Statement on Responses to Food Insecurity (2015), endorsed by 140 organizations, while recognizing that food charity operates under considerable constraints as the primary response to food insecurity, food banking is both ineffective and counterproductive. It fails to address the root cause which is income poverty and allows the misconception to continue that food insecurity is being addressed.

Significantly, the OSNPPH statement references nutrition and public health research which finds that 'adults living in food insecure households have poorer self-rated health, poorer mental and physical health, poorer oral health, greater stress, and are more likely to suffer from chronic conditions such as diabetes, high blood pressure, and anxiety' as well as the mental health impacts of food insecurity upon children (Vozoris and Tarasuk, 2003; Melchior et al, 2012; OSNPPH, 2015). Valerie Tarasuk has described food insecurity in Canada as a public health crisis (Riches and Tarasuk, 2014, p.45).

This concern is shared in the UK. A letter sent by public health professionals to the *British Medical Journal* and reported in the *New Statesman* argues that food poverty has 'all the signs of a public health emergency' and notes the comments of the vice-president of the UK's Faculty of Public Health regarding the reported rise of Victorian-era diseases caused by malnutrition, such as rickets and gout (McBain, 2015). Obviously food banks are not the cause of such conditions but if they are perceived to ensure sustainable access to adequate and nutritious foods they become part of the problem allowing governments to look the other way.

Martin Caraher, professor of food and health policy, and Sinéad Furey, lecturer in consumer studies, in a briefing paper for the UK's Food Research Collaboration add their voices pointing out the ineffectiveness of surplus food as a 'short-term band aid and not a way to address hunger, citizen's social rights to food or their nutritional needs' also noting that while 'there are short-term benefits for individuals', there is 'not enough research on the long-term impact of donations on citizen's health and food insecurity' (Grover, 2017; Caraher and Furey, 2017).

Consolidated research findings, also from a public health perspective, of a recent systematic review of the functions and efficacy of food banks in addressing food insecurity in high income countries, add weight to these criticisms. The study, conducted by researchers at Deakin University in Australia, shows that while 'food banks have an important role to play in providing immediate solutions to severe food deprivation, they are limited in their capacity to improve overall food security outcomes due to the limited provision of nutrient dense foods in insufficient amounts, especially from dairy, vegetables and fruits' (Bazerghi et al, 2016, p.732).

These findings reflect concerns about the nutritional adequacy of the surplus food on offer and the fact that food banks cannot guarantee meeting special dietary needs or cultural preferences. They are dependent on the food – processed, pre-packaged, canned, junk or fresh – which is supplied or donated.

Reflecting nutrition and public health research in Canada, the Deakin article cites studies from Australia and the USA which report that many experiencing long-term food deprivation, and having to rely on food banks, still remain food insecure. This is concerning given links between food insecurity and poor health (Bazerghi et al, 2016, p.738). Critically, part of the explanation lies 'in the extraordinary levels of vulnerability of those who seek food charity (i.e., it is a last resort)', but also 'the limited food assistance provided by this system' (see Riches and Tarasuk, 2014, p.45).

Organizational constraints – the limits of voluntarism

One could well argue that food banking successfully meets its primary aim of delivering short-term emergency food aid and as such is what makes a difference to people in need. However, after three decades of food banking in the OECD it is clear that short-term emergency food relief has become long-term food assistance. Not only are today's charitable food safety nets dependent upon Big Food and its corporate partners, they also could not survive without an extensive army of volunteer labour.

The fact that the vast majority of food bank staff are volunteers brings with it organizational headaches. This may well result in limited operating hours, create additional stress during holiday periods and weaken organizational capacity to mount continuous food bank drives. Attracting ever-increasing volumes of food is an ever-present onerous responsibility as food banks. Despite being oversupplied with bread and pastries food banks are always running out of food, processed, healthy, nutritious and fresh. This becomes particularly noticeable in the run up to Christmas when food bank shelves need to be fully stocked but also at other times of the year.

Such organizational demands may present volunteers with significant challenges. Recently in Germany the pressures on volunteers to process the enormous volume of food required to feed the needy on a weekly basis took could not cope. The Food Bank (Tafel) in Nuremburg, a city of half a million people, shut down unexpectedly in October 2016 because 'its 1600 volunteers were consistently overwhelmed by collecting and distributing 25 tonnes of food 5 days per week. City authorities could not offer assistance and consequently 6000 people went hungry for 10 days' (Tang, 2017). One could argue this is an isolated incident but the relentless demand for acquiring food and the logistical pressures and costs involved are a recurring theme in the history of food banking in the rich world.

Volunteers are at the interface between the person seeking food and the food on the shelves. Whatever the eligibility criteria adopted by food banks or food pantries the discretion of the volunteer is decisive and generally applicants will be well received. Yet shortage of food can only mean tighter rationing and smaller and less diverse food parcels being handed out but already limited to a few days supply with restrictions on return visits. It will also result in food banks closing early.

Unavoidably and unintentionally, with the best will in the food bank world, uncertainty about food supplies or its lack leads to the sharper application of eligibility criteria sorting out who is deserving: who gets to eat, who does not. Oliver Twist had a difficult time when he asked for more. Who does one turn to when 21st-century food banks run out of food or eligibility is denied when your wage packet prevents your eating or income assistance has been denied?

Reflections

Domestic hunger has for too long been accepted as an everyday reality and the legitimate responsibility of corporately managed food charity. It has been depoliticized. It is no longer a human rights issue demanding the attention of the State or as some would pejoratively term it, Big Government. We have instead come to depend on Big Philanthropy and the ascendant role of corporate food charity, locally, nationally and globally as solutions to food insecurity and domestic hunger in poor and rich countries alike. Food charity, the new safety net of the safety net, is now increasingly woven into the weft and warp of re-configured and irresponsible welfare states.

There is a moral claim to charity but not an enforceable human right. At this point charitable food banking, even when backed by Big Food and faith communities is not in a position adequately to alleviate hunger in the short term nor advance longer-term food security. Food banking's ineffectiveness in the wealthy OECD undermines its intentions to end hunger today or prevent hunger tomorrow and act in solidarity with the poor. Yet its 35-year march since crossing the US–Canada border in 1981 and instititionalization in today's first world signifies its corporately backed philanthropic success. As Linsey McGoey has written; 'philanthropy … is a model of giving that is not imperiled by its own ineffectiveness. Rather it thrives on it' (McGoey, 2015, p. 147). Meanwhile the indifferent State looks the other way.

7

CORPORATE FOOD CHARITY

False promises of solidarity

Assessing the effectiveness of food banking's claim of alleviating and reducing hunger and reducing food waste through the redistribution of surplus food is a necessary yardstick for determining whether food banks are a solution to food poverty. While food insecurity measures and data may be variously interpreted, even disputed, they are essential to guide informed public policy debate. However, in addition to evidence-based reasoning as to whether food banks successfully keep food on the plate for hungry people and corporate leftovers from the dump, the moral claim that corporately captured food philanthropy acts in solidarity with the poor demands attention.

A case in point: at the time of writing this chapter I had by chance been following the BBC World Service news on my tablet and was surprised by an advertisement which popped up from Feeding America and the non-profit US Advertising Council soliciting donations (I was at home in Canada). The advertisement reads

> **'Together we can fill their fridge' – Donate Now –**
> **'Together we're Feeding America'**
>
> *(BBC, 11.8.17)*

I am unsure why this request is beamed across the 49th parallel but it surely was an invitation to donate in solidarity. Then thinking about corporate food charity I wondered solidarity with whom.

It took me back to the 1980s and the early days of food banks, the subsequent rise of the food bank nations in the rich world and who is benefiting and why when 'left over' food is being served to 'left over' people, as Elizabeth Dowler once expressed it. What are the moral choices confronting society and the State – the public interest – as to whether corporate food charity should be considered a normal and everyday response to widespread domestic hunger; and what are the ethical considerations that individuals volunteering and working within the now globally institutionalized food banking industry might wish to think about.

Solidarity: the moral imperative

One way to engage with these matters of social values and human rights is through the lens of solidarity particularly as this commitment is one which drives the global food bank movement. The moral imperative to feed hungry people is fundamental to charitable food banking. It expresses an idea central to solidarity, a unity of interests, purpose and action that must be taken to support those unable to feed themselves and their families. It is about practical compassion: caring for the other; walking in another's shoes; there but for the grace of God go I; we are all in this together; giving to strangers; and standing shoulder to shoulder with the socially excluded. Individuals and faith-based communities doing their bit by donating food, assisting in food drives and volunteering at food banks are expressing a commitment to an idea of solidarity. This commitment could also be said to motivate transnational conglomerates and local companies contributing from a sense of corporate social responsibility such as Big Food, the food retailing and services industry and philanthropic foundations which supply food and funds and whose representatives may sit on food bank boards managing the newly established food safety nets.

For individuals and corporate actors the measure of solidarity becomes the daily task of rescuing increasing supplies of wasted food heading for the landfill, creating bigger and better food banks, building innovative food distribution hubs so as to feed people in need. The implicit message conveyed of solidarity is one of standing shoulder to shoulder with the poor and socially excluded through making sure no goes hungry.

Solidarity in Europe

This message is certainly expressed by the values informing the mission of the European Federation of Food Banks. Along with giving (free of charge), sharing (fairly and without discrimination) and fighting food waste, solidarity is a principal value informing FEBA's philosophy and work. Its call for solidarity means 'distributing (food) to charitable agencies that help the most deprived people in Europe, fostering volunteering and promoting social solidarities' (FEBA, 2017f). It is foundational to FEBA's mission and in tune with European social and cultural norms.

Solidarity is a prominent value central to the EU's origins and its history as well to European social policy. Catherine Barnard, Cambridge professor of European Union law, notes that the word solidarity is found in several key EU constitutional documents including the Preambles of the Constitutional Treaty, the Lisbon Treaty and the Charter of Fundamental Rights 2000. It is also expressed in the EU Commission's thinking about a new social policy vision for Europe including the 2008 document *The Renewed Social Agenda: 'Opportunities, Access and Solidarity'*

(Barnard, 2014). Yet Barnard critically asks whether the concept of solidarity is 'employed as much for its rhetorical value as its substance' (p.74).

Of particular interest in the context of feeding the hungry, Barnard, referencing Alain Supiot (writing about work and EU Labour Law), also observes 'that "such solidarity must not be conceived of merely in terms of a response to individual need" since that would lead to a shift from the welfare state to a "charity state" but rather as expressed in the Solidarity Title of the Charter of Fundamental Rights (2000) as implying "also an active role in terms of, for example, social assistance and the provision of health care"' (ibid, p.102; Supiot, 2001).

Historically a commitment to solidarity has underpinned the universality of the Nordic welfare states though the arrival of food banks in Finland and more recently in Denmark, Norway and Sweden suggest some fraying at the edge. Similarly in food bank Germany solidarity as professor Stefan Selke ironically points out is a mark of 'the German constitution's criterion of unconditional help in the context of an institutionalised system of solidarity in a welfare state' (Selke, 2013). Universality and publicly funded social safety nets are expressions of solidarity and social rights which are in tension with the practical compassion of corporately organized food banking and the newly emerging food safety nets of the charity state.

Yet in Catholic Spain, Karlos Pérez de Armiño when considering the appeal of solidarity and human values within the food banking movement, writes that the Spanish Federation of Food Banks (FESBAL) claims to have 'organised a new concept of "solidarity", linking industries with charities' thereby 'making a commitment to a more prosperous world' (FESBAL, 2013). This commitment likewise involves a civic moral obligation for each citizen to become involved in anti-hunger activities through the formation of local 'solidarity networks'; and involves the struggle against food waste and the fight for social justice (Pérez de Armiño, 2014). In other words the concept of solidarity is understood as a universal value, perhaps with particular resonance in Catholic countries where it is rooted not only in religious teaching and family values but also in political discourse and the commitment to equality and social justice.

These commitments, rhetorical or not, express conflicting views about the meaning of solidarity. The tension however between practical compassion, the moral imperative motivating food banks to feed hungry people, and social justice, the human right to feed oneself with choice and dignity, remains. Nor is it likely to be resolved by the corporate social responsibility of Big Food and its partnership with the food bank industry considered by FESBAL to be a new concept of solidarity. Indeed in the USA this partnership has been called an unholy alliance (Fisher, 2017)

Big Food's corporate social responsibility

Corporate social responsibility is to be found at all levels of the food industry, globally, nationally and locally. It carries a message of partnership and solidarity through building alliances with food charity organizations, cementing the idea of Foodbanks Inc.

The transnational agribusiness Cargill is an exemplar of Big Food's alliance with food charity. For more than 25 years it has been in partnership with Feeding America. It sees its commitment as one of closing the meal gap, reducing food waste and helping to 'nourish millions in need', in other words to 'provide better access to food in local communities, while working to find long-term solutions to hunger' (Cargill, 2015). Cargill was also a founding member of the Global Foodbanking Network in 2006.

Across the rich world Cargill has supported the establishment of food bank networks in Mexico, their expansion in Canada and likewise in Hungary, Poland and Spain. In Europe it has also worked closely with FEBA providing funding, product donations and with staff offering time and management expertise as volunteers.

The UK's FareShare has also benefited from Cargill's support in the founding of a new food bank in Liverpool. In Europe another prime example is the call for international solidarity by the Carrefour Foundation acting on behalf of the French food conglomerate Carrefour, in partnership with FEBA and the GFN to coordinate an international operation to collect food products to give to food banks in eight countries (CF, 2013)

Importantly, Oxfam's briefing paper *Behind the Brands* (2013) brings attention to the UN's recognition in 2011 of 'the vast human rights impacts of businesses'. This has led Oxfam to endorse 'a detailed set of responsibilities to all companies'. Global business is now required to put in place policies and procedures to identify human rights issues across their supply chains and engage with all stakeholders, including governments to address them (Oxfam, 2013, p.16). Notably the briefing paper is focused on food justice and the role played by the 'Big 10' food and beverage companies (Box 7.1), the most visible corporate actors within the global food system who wield immense power.

BOX 7.1 'BIG 10' FOOD AND BEVERAGE COMPANIES, 2013

Associated British Foods (ABF), Coca-Cola, Danone, General Mills, Kellogg, Mars, Mondelez International (previously Kraft Foods), Nestlé, PepsiCo, Unilever

Source: Oxfam, 2013

Oxfam claims that the focus of the food and beverage industry giants has not always been to make 'sure everyone always has enough nutritious food to eat'. Rather it argues that over the past century it 'has used cheap land and labor to produce the least expensive products possible – often of low nutritional value – while maximizing profits. As a consequences water resources have been depleted, greenhouse gas emissions have risen and labor has been exploited, yet the food industry and is shareholders have prospered.' In the meantime, in addition to hunger, Oxfam notes that global epidemics of diabetes and obesity are linked to the food industry's production of 'junk food' and sugary beverages (see Oxfam, 2013, pp.5–6).

To counter such adverse consequences Oxfam's strategy has been to promote a 'Behind the Brands' campaign working with the 'Big 10' to change the way the Big Food companies do business across the global food system particularly in low income countries. Working with the Big 10 Oxfam has introduced a Food Companies Scorecard assessing each company's performance in relation to land, water and climate issues; the rights and conditions of women, farmers and workers; and transparency. As CEOs cite, there are 'a variety of reasons why it makes business sense for them to be attentive to corporate responsibility, including meeting ethical and philanthropic responsibilities, developing and maintaining legitimacy and reputational capital, and building stronger relationships with stakeholders' (Oxfam, 2013, p.18).

In the rich OECD world these stakeholders include the 'Big 10' global food brands which are integral to and active partners or supporters of the charitable food bank industry in the rich world redistributing huge amounts of their surplus food to feed hungry people and reduce food waste. In reflecting the moral imperative to feed hungry people they are presented as corporate examples of good will. The question, however, is whether these are acts of solidarity and if so who is benefiting. Is it Big Food's solidarity with its partners in charitable food banking we are being asked to support or those too poor to feed themselves and their families?

'Uncritical' solidarity: hidden functions, false promises

Big Food's corporate capture of food banking coupled with the indifference of austerity minded governments which have lost their moral compass questions the idea of solidarity. This invites consideration of the hidden functions of institutionalized food charity and food safety nets. As Karlos Pérez de Armiño writing about Spanish food banking explains it is rather 'a way of uncritical solidarity'. It fails to address critical concerns posed by anti-poverty and social justice organizations when 'no analysis can be found on the causes of hunger or poverty, nor any criticisms of the institutions, their public policies in the framework of the crisis, or the economic system' (Pérez de Armiño, 2014, p.144). Certainly food banking and its corporate sponsors appear to promise solidarity with the poor. Yet there is little evidence that this translates into advancing social and justice, let alone the right to food.

Will corporate social responsibility lead to solutions to food poverty grounded in human rights and economic and social justice? Is the building and institutionalization of charitable food safety nets, dependent on the surplus food, funding and expertise of the food industry, an effective solution or a distraction from the causes of food poverty and food waste? Is the process of building 'charitable' food safety nets taking the politics out of hunger and disarming us all from the urgent challenges and necessity for community and social action? Or is Big Food's partnership with the food bank nation rather a case of corporate social investment (CSI) leaving the neo-liberal *status quo* preserved? More likely the answers to these questions uncover the false promises of solidarity.

Corporate social investment

Historically and in the early days of food banking it was the charities seeking the food retailers left over food. Today, it is a different scenario, with Big Food and its industry partners sitting firmly in the global food bank saddle holding the reins and diverting endless amounts of surplus food from landfills to feed the hungry. This level and degree of the food sector's commitment to corporate social responsibility, 'doing well by doing good', is surely to be applauded. Yet when thinking about the food industry's role in the development of food safety nets in high income countries there is a flip side to corporate 'do gooding'. Perhaps it is commercial motivation or CSI which drives the agenda.

This question was recently posed by the Ontario Society of Nutrition Professionals in Public Health when commenting on the significant control and influence which CSR exerts over food charity and food insecurity in Canada's largest province. It recognizes that corporations participate as board members for national and provincial food charity organizations and provide significant food and monetary donations. Yet, it also comments that as market research has shown companies acting in this way 'build brand loyalty, attract new customers, drive word of mouth advertising and grow revenue. They also benefit from donating unsaleable food by avoiding landfill disposal fees' (OSNPPH, 2015 p.3). In other words CSR functions primarily as a form of CSI.

If the idea of solidarity is standing shoulder to shoulder with the poor we need constantly to be unmasking the hidden functions of the corporate capture of food banking and food safety nets – raising for public debate the often unaddressed moral certitudes informing its purposes and modes of operation. Who stands to benefit and why from Big Food's take over of food charity from public welfare?

Whose shame should it be when fellow citizens are on the receiving end of food hand outs? Why has it been left to the food industry and its charitable food bank partners to manage hunger in the process disciplining the poor and creating a secondary system of welfare? Who is excluded, continuing to be on the outside looking in? Where does accountability lie? Morally, is feeding wasted and surplus food to people who are surplus to the labour market the best we can do in today's affluent western style wealthy world? This question invites revisiting the practice of corporate tax incentives as a form of corporate social investment.

Corporate tax incentives

It should be no surprise that tax deductions and exemptions aimed at discouraging food waste and encouraging charitable food donations are increasingly promoted. Why not? They boost the food industry's sense of corporate social responsibility, with the added benefit of padding a company's profitability. Self interest serving the welfare of all.

In recent years, Food Banks Canada has lobbied the former Federal Government to support Ottawa's plan to stimulate the national charitable sector by introducing a tax incentive plan for food donations (FBC, 2012b); and more recently has backed a proposed municipal zero waste movement also intent on lobbying the Federal Government to legislate tax incentives to reduce food waste. Food donation tax credits for farmers have been introduced at the provincial level in Quebec, Ontario and most recently British Columbia. While certainly benefiting the farmers, as Robin Roff, a researcher with the Canadian Centre for Policy Alternatives, has argued while such tax credits appear to be 'a creative "win-win" solution' ... 'food banks are not the solution to food insecurity, either in the short or long term and should not be treated as a de facto public policy' (Roff, 2016)

In Europe a comparative study conducted by Deloitte on behalf of the European Economic and Social Committee shows, not unsurprisingly, that 'fiscal incentives through tax credits and tax deductions encourage food donations'. It notes that 'in France 60% and in Spain 35% of the value of donated food can be claimed as a corporate tax credit, meaning that food donors are able to deduct that percentage of the value of the donated food from the corporate tax on their revenue' (Bio/Deloitte, 2014).

While there is no common EU policy regarding food donations, in Germany, Greece, Hungary, Italy, Poland and Portugal, VAT is not imposed on food donations and they are also treated as a deductible tax expense thereby reducing a company's taxable income. However the value of the tax break varies according to the member state. In Belgium, Denmark, France, Spain, Sweden and the UK there is a mix of policies (Bio/Deloitte, 2014). Nevertheless whatever the food donation or food waste tax exemption policies the food processors and retailers will always be better off

French law is now attracting wide attention in the EU since it has banned supermarkets (over 400 sq metres) from dumping food in backstreet bins or landfill, requiring them instead to sign donation contracts with food banks and charities, a move supported by Carrefour, France's largest supermarket chain (Chrisafis, 2016). A legislative stick and carrot approach is being used to abolish food waste and redefine it as surplus food necessary for feeding hungry people. The French food business option is now either face a hefty fine (Euro3750) or benefit from the 60% corporate tax credit. The latter choice can only improve supermarket profit margins and further solidify the food bank safety net, the new legally approved checkpoint at the end of the food bank benefit chain.

Harvard Food Law and Policy Clinic

None of the recent food donation tax incentive initiatives in Europe and Canada are likely of much interest in the USA where the country's established public–private food safety net has benefited from more than fifty years of federal legislation encouraging food donations including the Tax Reform Act (1976), the Good Samaritan Act (1986), the Bill Emerson Good Samaritan Act (1996) and the Food Donation Act (2008). What I find surprising and discouraging is the role played by the Harvard Food Law and Policy Clinic

In a recent briefing paper promoting food donations and explaining the intricacies of federal tax incentive legislation and its benefits for business, Harvard's Food Law and Policy Clinic (FLPC) advises that 'federal tax incentives provide financial incentives that make food donations more cost effective and economically beneficial' and 'have been extraordinarily successful' (FLPC, 2016). As evidence, the FLPC reports that following the 2005 temporary expansion of coverage to include more businesses, food donations increased by 137% in 2006, with Congress making the expansion permanent in 2015 (FLPC, 2016). The intention has not only been to increase donations for the purpose of preventing food waste, but explicitly to make the case that such actions are good for business profitability, and doubtless for corporate social investment.

The FLPC briefing paper notes that only some US states augment federal tax incentive food donation law with legislation of their own. They do not include Massachusetts, home to Harvard and its Law School. For that reason the FLPC is mounting an awareness and advocacy campaign in its own state focused on promoting food donations to feed hungry people as a strategy for reducing food waste. This shows remarkable commitment by the best and the brightest to further strengthen the institutionalization of the US corporate food bank safety net. As such it stands out as a prime example of 'uncritical' solidarity and of the false promises of corporate food charity at work.

The critical question is why devote so much high legal intellect and ability to advancing ineffective food donation remedies to hunger and food waste when their causes lie in dysfunctional food systems and broken social safety nets of which corporate food banks are symptoms and failing extensions. Every tax deductible dollar supporting a corporate donation of food is a dollar lost to public revenue. More to the point, from the perspective of solidarity, should we not be asking whether leaving the management of hunger to Big Food and under-resourced and fragmented charity is benefiting or punishing the poor.

Shaming the hungry, regulating the poor

In the fall of 2014 I had the opportunity to revisit Glasgow where I had spent time in 1965 in the Gorbals as a social policy student learning about poverty. This time it was to attend the annual conference of Nourish Scotland, the pre-eminent NGO campaigning for food justice in Scotland. The focus of the conference was exploring the global food system so as to think more clearly about the way forward for Scotland taking steps towards an ethical food policy. It was attended by 200 people from various organizations including food producers, policy makers, environmentalists, public health staff, academics, human rights and civil society activists (NS, 2014). It was a very successful event, challenging and highly educational.

The conference was also the first time after nearly three decades of research and writing about food charity that I heard a senior government official, the head of Scotland's Food, Drink and Rural Affairs Office, publicly state how ashamed he was that food banks had come to Scotland. Three years later Elli Kontorradvis of Nourish Scotland believes it is 'now normal for politicians in Scotland to say they are ashamed of the need for food banks' (EK – email communication28.9.17). Shame after all is not only experienced by those standing in food bank line ups.

This is not to suggest that shaming the hungry is a deliberate policy enforced by food banks, but it is a hidden, unintended and unavoidable consequence of charitable food provisioning. It is the stigma and experience of being made to feel 'other' when having to beg for food, with little or no choice in the food being handed to you (Selke, 2013; Garthwaite, 2016; Lambie-Mumford, 2017). It is a loss of human rights and dignity. It is an expression of a sense of powerlessness. Not unsurprisingly, as studies in New Zealand, the Netherlands, Germany and the UK have shown, shame, or the fear of it, is first experienced by hungry people too embarrassed ever to set foot in a food bank. Some delay, some do not go at all (McPherson, 2006; Placido and Rietberg, n.d.; van der Horst et al, 2014). People have too much pride to step inside.

Of course for many there is no alternative, including mothers with children to feed (Tarasuk and Beaton, 1999) or those whose credit has run out, facing huge debts, fixed but unpayable rent and fuel bills to say nothing of the everyday costs of normal living and of rising food prices. However cheap food may appear to be, food is simply unaffordable for those who are living precarious lives trying to get by on 'zero hours contracts', inadequate minimum wages and welfare benefits or who have been sanctioned and denied eligibility for financial assistance in the short and long term. Food is also the elastic item in the household budget. It is what you cut down on or go without whether you are working poor, living on a First Nation reserve, a refugee in a hostel or homeless and living on the street. In such circumstances many people are forced to set their personal dignity aside and cross the food bank threshold.

The 'dark side' of food banking

In the Netherlands as Hilje van der Horst and fellow researchers remind us there is a 'dark side' to food banks (as surely elsewhere in the rich world) bound up in the emotional responses of those for whom they are a place of last resort. The question explored was whether 'food banks generate interactions of charitable giving that may be harmful to the self-esteem of receivers' (van der Horst et al, 2014).

Whilst gratitude was expected by volunteers (see also Tarasuk and Eakin, 2003, 2005) and expressed by some recipients, the study tells that the most prominent emotion experienced by those on the receiving end of food parcels were feelings of shame. Shame about the content and the quality of the food being provided (on or beyond the expiry date; spoiled food; high in fat and sugar) and of what one was expected to eat ('otherwise given to pigs'); when first visiting the food bank (which dampens somewhat over time) and interactions with volunteers who expect at least a modicum of gratitude as part of the social rules and norms of food bank behaviour; and about the stigmatization of being a food bank user and the guilty feeling that you have to address the issue of personal fault (van der Horst et al, 2014).

As the Dutch researchers rightly observe 'the experience of needing to ask for help with a very basic human need, the need for food, (is) exceptionally degrading' (van der Horst et al, 2014). Canadian Elaine Power has also written how 'people find the experience of using a food bank stigmatizing, humiliating and degrading, no matter how kind the food bank staff and clients are' (Power, 2011).

In Germany, sociology professor Stefan Selke's research shows that 'those who use food banks are being placed into a vulnerable situation with both physical and psychological stress factors'. The physical causes of stress 'include queuing in public and long waiting times'. The psychological stress factors repeatedly named were 'fear (of being seen by others), embarrassment, restrictions (no choice, no right to complain) and dutiful responses (statements of gratitude, rituals of submission). However the main stress factor is shame. There is a fundamental conflict in that. Voluntary involvement is seen, on the one hand, as socially desirable while at the same time many emotionally stressful experiences of shame and denial are accepted and socially permitted by foodbank volunteers' (Selke, 2013, p.2).

Indeed feelings of shame get to the heart of the moral and ethical issues of distributing 'left over' food to 'left behind' people dressed up as a solution both to food waste and food insecurity. While the action is well intentioned it denies choice and human dignity – the right to choose the food one eats and provides for one's family. Given that we live in prosperous market economies, the normal and customary way is to go enter a store and purchase the food of your choice.

Managing the hungry, disciplining the poor

Food banks and charitable food provisioning as they have evolved over time are not unaware of these issues and certainly ways of providing increasing volumes of fresh and healthy foods and offering as much choice as possible are important developments. However, the food on offer is only as sufficient, nutritious, healthy and as fresh as the volumes of surplus food received from the food industry, individual donations and whatever can be purchased including canned and processed products. Food bank supplies are always limited, never guaranteed and always running short, otherwise corporate and volunteer propelled food drives would not have become so frequent a moral obligation demanded of communities everywhere

The point is this: each day when charitable food banks, food pantries and meal programmes open for business, they assume responsibility for overseeing and administering what used to be called, in the English speaking world, parish or poor law relief. At that moment, food banks and their volunteers accept responsibility for managing the hungry and regulating, some would say disciplining, the poor. For practical reasons limited food supplies coupled with significant demand necessitate eligibility criteria being determined and applied – who gets to receive food or not (who is genuinely in crisis, who is not; who is deserving, who is undeserving); as well as rationing the surplus food on offer – what kind, how much, for how many days (will repeat visits be allowed). Many times over, no doubt, extremely difficult decisions will have to be made by frontline volunteers particularly when food bank stocks are low. The food assistance you receive will depend on their discretion.

The charitable food safety net as it determines who is eligible (or not) and hands out food (or not) will be familiar to those in the food line ups, especially those who have had to deal face to face with similar eligibility requirements and the limits placed on income benefits by state welfare authorities. They will have had to deal with officials (complying with welfare legislation) who have rejected or delayed their claims for financial assistance, refused to increase benefits or sanctioned those unable or unwilling to accept or look for work at sub-poverty wages. Claimants of financial assistance have always been dependent on the discretion of the welfare official. They now are dependent on the good will and discretion of the army of food bank volunteers who are the outsourced unpaid workers of the newly configured corporate food charity state.

The corporate food charity state

Those who enter the food bank door may arrive of their own accord or clutching a referral form granting eligibility signed by a responsible professional or community leader. Compared to the welfare office they will likely receive a kinder, more understanding welcome, and the immediate food assistance they need. However they should be under no illusion that they have changed one welfare system for another. Charitable food banks are no more than an extension and unaccountable secondary tier of the welfare system, assisting in managing the hungry and regulating the poor on behalf of the indifferent state unwilling to invest public funds in the social safety net.

It can be argued that well-intentioned proponents of innovation and social enterprise have worked to take the sting of shame and humiliation out of the food bank experience by offering a degree of choice to people on low incomes to shop in social supermarkets or community food stores (well-established across Europe) or use voucher schemes at farmers' markets. Restaurants or cafes may offer free or 'only-pay-what-you-can' meals (subsidized by other patrons) and charitable meal exchange programmes have expanded. These are promoted as a triple 'win–win–win' solution to food poverty (Graslie, 2013). Yet the stigma still remains.

Furthermore along with food banks such initiatives are evidence of the development of a parallel system of food charity (Kessl, 2016), a food safety net comprising *secondary* welfare systems and *secondary* food markets for *secondary* consumers: hallmarks of the arrival of the corporate food charity state. This is not solidarity in action which ensures the human right to adequate food. People still lack the money, and continue to experience the stigma of being unable to shop for food like anybody else in normal and customary ways in mainstream supermarkets and grocery stores.

Taking the politics out of hunger: who benefits and why

When it comes to confronting the reasons why growing numbers of vulnerable people in high income OECD member states are having to resort to food banks dependent on waste food diverted from landfills, the hard truth is that governments have looked away or conveniently let it happen with a subsidy here or a tax break there. This is especially the case in the USA which has backed the development of Foodbanks Inc with a long history of legislation. The corporate capture of charitable food banking has masked the upstream structural causes of domestic hunger and food waste: broken social safety nets and a dysfunctional global food system. Only in Brexit divided Britain it seems are food banks a topic for political debate. Elsewhere in the wealthy world of food bank nations their corporate charitable sponsors have succeeded in taking the politics out of domestic hunger which has been socially constructed as a matter for charity and wasted food, a technical problem which can be fixed.

As welfare states have retreated there can be little doubt that philanthropy, big and small, has privatized emergency food assistance and institutionalized food safety nets at the expense of publicly funded income assistance. The problem is that despite evidence to the contrary, food banks are publicly perceived as expressions of solidarity with the poor and effective 'win-win' solutions for reducing food waste and domestic hunger. In the meantime corporate food banking's 'ostensible effectiveness offers national governments a way to "outsource" the political risk of domestic hunger' (Silvasti and Riches, 2014, p.202), evidence that governments have lost their moral compass.

De-politicization of hunger

Foodbanking has taken politics out of hunger allowing the public and governments to believe that the income poor are well looked after by corporately backed food charity. In the Gilded Age in which we live this is a dream come true for established power elites preaching the mantra of neo-liberalism: economic growth; jobs (any jobs); small government; deregulation; lower taxes; and the downloading of social protection to the community with the assumption of philanthropic support. As this ideology has unfolded aided by 'the corporate self promotion of their food charity efforts and media coverage, food charity has proved to be an acceptable and appropriate response to food insecurity' (OSNPPH, 2015).

Institutionalized food banking has led to the creation of secondary welfare systems and food markets for secondary consumers and the establishment of a parallel system, or a pre-welfare system of food charity – a stop gap safety net for the underfunded social safety net. This has enabled governments to impose strictly enforced welfare eligibility criteria restricting or denying access to financial assistance for those able bodied people lacking attachment to the labour market, claiming they are saving taxpayers the costs of unnecessary income security and social welfare spending.

In the process the indifferent State absolves itself from responsibility for the basic food needs of vulnerable citizens. The social contract is broken. Governments are no longer publicly accountable: domestic hunger has been expunged from the political agenda. There is no worry. Big Food and its industry partners acting from a sense of solidarity and corporate social responsibility will play philanthropic backstop to food banks, advance food security and serve as the new protectors of the hungry poor. Or do they?

Who benefits and why

In light of the promise of solidarity who really benefits and why from the corporate capture of food banking and the re-cycling of its surplus food to feed hungry people? Given the upstream problems of broken social safety nets and dysfunctional over-supplying food systems, to say nothing of the limits placed on food banks by their dependence on the happenstance of the amount and quality of surplus food they receive. Leaving aside the moral and ethical issues of using 'left over' food to feed 'left behind' people, there is scant evidence the benefits are primarily experienced by the millions using food banks and certainly not by those too ashamed even to enter the door and beg for food. The fact is too many cannot even afford to put cheap food on the table.

In the rich world food banking's main beneficiaries are Big Food and the State. Within a culture of philanthrocapitalism, Big Food conglomerates, agri-business, producers, supermarkets and restaurants acting from a sense of solidarity and corporate social responsibility profit literally and figuratively from demonstrating good will, a sound business sense and competitive branding. It is a form of corporate social investment. In their partnering with food bank associations they are seen to have stepped into the breach with public support on their side. Over time Big Food, Big Philanthropy and Food Banks have become the gate keepers of today's charitable food safety nets in the OECD. The corporately captured food bank nation has developed a life of its own and the model of US food assistance is being exported to the rich world. This must be good for business.

Governments and ourselves as taxpayers and consumers are likewise major beneficiaries. The *uncritical* or misleading sense of solidarity which has informed the rise of food bank nations enables the State to rest assured the poor are being fed and no concern of theirs. Public revenues do not have to be increased and they can keep on promising the neo-liberal mantra of ever lower taxes. Minimum wage rates can be held down; living wages contested and income assistance cut or denied. Governments at all levels appear to feel no shame, indeed food banks are lauded in the USA, though other national governments will tend to be more Janus like – looking both ways, turning a blind eye, but passing food safety, food waste and tax exemption (foregone revenue) legislation to help the cause.

While some politicians may be ambivalent, few will be bold enough to criticize Big Food's corporate global capture of food banking and the spread of US-style charitable food safety nets to other OECD member states. Why interfere when food banks are seen to be regulating the hungry and the poor, and, as a double act, purportedly providing a 'win-win' solution to food waste.

It also needs to be emphasized that the work of food banks depends upon thousands of donors and volunteers providing free labour. To what extent might they be beneficiaries? Despite the contradictions and ambiguities in their roles, volunteers actively express their practical compassion by working on the frontlines to alleviate hunger by delivering food parcels to those in need. While research shows that volunteers hold different views about the deservingness or otherwise of food bank recipients, nevertheless they are acting on their beliefs and doing their bit. The media and online food bank publicity will often portray volunteers' strong sense of camaraderie and community responsibility with pictures of them always smiling.

Yet, as indeed some are, they should be concerned about food banking's ineffectiveness and its claims of solidarity with the poor. As observed twenty years ago the moral imperative to feed hungry people acts rather as a moral safety valve (Poppendieck, 1998; Berg, 2008) allowing us to believe we have done our bit and yet the structural and primary determinants of food poverty are left untouched.

The global spread of US style food banking has not demonstrated it is part of an effective solution to hunger and food insecurity, but part of the problem. Governments have taken advantage. Neglecting income security and publicly funded social safety nets in favour of food assistance dependent on surplus food to feed hungry people is not only ineffective it denies choice and undermines human dignity. It distracts government's attention from public policy informed by the human right to adequate food.

Reflections

The false promises of solidarity are evident in the corporate capture of food banks with their dependence on the leftover food products of Big Food and supermarkets. Enabling indifferent governments to look the other way, institutionalized food charity merely permits the continuing imposition of neoliberal welfare policies designed to manage and regulate the poor. Welfare benefits remain so low that hungry people are pushed into low wage work, even assuming it is available, which itself forces people with a strong work ethic to the food bank shelves. Charitable food safety nets serving as safety nets for underfunded and broken systems of income security mask the underlying causes of domestic hunger and distract attention from the human right to adequate food. Corporate food banking promises solidarity delivering more of the same further entrenching food inequalities.

Whilst the global charitable food banking industry claims to have a *'win-win'* solution for reducing food waste and domestic hunger, this is an empty and misleading promise. It perpetuates myths about the capacity and the commitment of corporate charity to solidarity and social rights all the while undermining government accountability and effective public policy. Rather than acting in solidarity with the poor by drawing attention to income poverty and inequality the hidden functions of corporate food banking obscure questions of structural causes and who is really benefiting and why. It allows the State to become a bystander to widening income inequality and food injustice.

Whilst state responsibility for addressing domestic hunger is off loaded and downloaded to corporate food charity, food banking has proved to be no more than 'a way of uncritical solidarity'. In light of the ineffectiveness and the moral dilemmas and ethical issues confronting the food bank nation project we need to ask whether diverting wasted food from landfills to feed millions of left behind people is really the best that wealthy OECD member states can do as a solution to domestic hunger and food poverty.

In moving 'beyond surpluses' it is necessary to change the conversation from *uncritical* to *collective solidarity* informed by the human right to adequate food.

PART III
Rights talk and public policy

8

COLLECTIVE SOLIDARITY AND THE RIGHT TO FOOD

Moral, legal and political obligations

In 2003 I had the good fortune to be invited by Stuart Clark, Senior Policy Advisor of the Canadian Food Grains Bank (CFGB), on behalf of the FAO Council's newly established Intergovernmental Working Group (IWG) based in Rome, to prepare a case study about the right to food in Canada. The invitation was a surprise given the FAO and CFGB focus on hunger and food security in the Global South. The IWG had been charged with developing national guidelines for implementing the right to food in low income countries and emerging economies. It had solicited case studies from Brazil, India, Mali, South Africa and Uganda and for comparative purposes wished to include one high income UN member state, thus the invitation to Canada.

World Food Summits and Right to Food Guidelines

The backdrop to the FAO's new rights based approach can be traced to the 2002 'World Food Summit: Five Years Later' (FAO, 2002) which endorsed the previous declaration of the 1996 World Food Summit at which Heads of State and Governments had affirmed 'the right of everyone to have access to safe and nutritious food, consistent with the right to adequate food and the fundamental right of everyone to be free from hunger' (Eide, 1998). In this context the specific intent of the IWG case studies was to provide background material to advise and support governments with developing national framework legislation and action plans for addressing malnutrition, hunger and food insecurity informed by rights-based approaches.

In other words states should not only address economic development and supply side food production matters (the responsibility of the FAO's Economic Division) but focus on a more comprehensive human rights framework particularly directing attention to issues of food access and distribution in the poor world. The shift to a rights-based approach was long overdue given that the vast majority of UN member states, including those invited to conduct case studies, had already ratified the *International Covenant on Economic, Social and Cultural Rights (ICESCR)*, including the right to food, which came into force in 1976.

The national case studies sought evidence for the right to food in practice; its constitutional entrenchment and justiciability; where it might be found in national legislation and public policy; and whether the right to food had legs in creating access to food. The Canadian study benefited from expert cross-disciplinary contributions in law, agriculture and food policy; nutrition and public health and social policy (Riches et al, 2004). As a result of this process and wider consultations, including significant opposition by powerful states and non-state actors, the FAO approved and published the *Voluntary Guidelines to support the progressive realisation of the right to food in the context of national food security* (FAO-VGs, 2005).

Global South to Global North

Directed at the Global South the Right to Food Guidelines (nineteen in all) are, as Katharine Cresswell Riol writes, exactly that: 'a set out of guidelines' (Cresswell Riol, 2017), and voluntary at that. They are designed to assist national governments translate international right to food obligations into national framework law, policies and practice.

Notably the guidelines have been described by leading international right to food experts and activists as being 'better than nothing' and 'as a "compromise paragraph" reduced to such by influential states who made efforts to substantially attenuate the idea of a code of conduct' (Cresswell Riol, 2017 quoting Germann, 2009).

Indeed Jean Ziegler, the first UN Special Rapporteur on the Right to Food expressed disappointment at the weakness of 'the language regarding the political and legal obligations implied by the right to adequate food' and 'the non-committal language, especially in regard to accountability' (Cresswell Riol, 2017; UN, 2004).

These views are shared by Hilal Elver, the current UN Special Rapporteur. In remarks to the UN Committee on Food Security (CFS) in early 2017 she commented that right to food language was ignored in the SDGs, and 'there was no mention of human rights in the operational provisions' of the 2015 Paris Climate Agreement (Elver, 2017), noting concerted opposition by Big Agro to the inclusion of right to food thinking in the Climate Accord. As for the SDG's Goal No 2, she argued 'it would have been more appropriate to affirm the right to food if the world really wants to **"end hunger, achieve food security and improve nutrition and promote sustainable agriculture"**' (bold in text). Gloomily, Elver sees support for 'the right to food approach as an uphill battle everywhere', driven by 'chauvinist nationalism and market driven thinking' (Elver, 2017, p.3). Charity remains the preferred big business response to hunger.

Yet it has to be said that during his mandate Olivier De Schutter the previous UN Special Rapporteur on his many country missions fearlessly promoted the

Right to Food Guidelines drawing on its first priority of 'democracy, good governance, human rights and the rule of law'. In Canada he reminded us that the human right to adequate food applies as much to the affluent Global North where neoliberalism and austerity driven governments have increasingly ignored the basic needs of vulnerable peoples, particularly those unable to feed themselves and their families, by relying on charitable food banking. This assessment surely holds true in the OECD food bank nations, especially the United States, the birthplace of the modern food banking movement, where corporate food charity has become institutionalized.

As Andy Fisher reminds us 'few anti-hunger groups in the United States, with notable exceptions, utilize a human rights discourse'. He quotes Manalo Le Claire, executive director of California Food Policy Advocates who states that '"we found over time that the right to food language works better in an international context than in a domestic context. Legislators in Sacramento believe a right equals a mandate which equals money"' (Fisher, 2017, pp.35–36). Precisely so but surely misses the point. The right to food makes the case for feeding oneself with human dignity in the Global North and not shying away from arguments about fair income distribution likewise a crucial issue in the Global South. Food is a political matter.

Yet, in the USA, as Fisher writes, the preferred approach, is reliance on the moral imperative of ending hunger and focusing on food as a human need not a right (Fisher, 2017, p.36). This willingness to dismiss or at least sidestep the right to food as central to the conversation about ending domestic hunger in high income countries is problematic. Hilal Elver's gloomy prognostication about the right to food and continuing reliance on business driven charity, questions food banks acting in solidarity with those dependent on eating wasted and surplus food. How we understand the concept of solidarity is crucial to understanding what the right to food means, what it does not and how it might inform food and social justice in the wealthy OECD.

Thinking about solidarity and food banking

It is necessary to move beyond conventional thinking about solidarity as raised in the previous chapter and consider where the concepts of 'critical' and 'collective' solidarity as bedrock values informing the right to food might lead. There are a number of reasons for this. Firstly that solidarity in European political culture and in EU social policy, as Barnard has suggested (see Chapter 7), may be understood as a rhetorical rallying call and response to individual need thereby leading to the 'charity' state, an approach which I would suggest has been adopted by the European food bank movement. Solidarity has even been employed by the Global Foodbank Network in building its worldwide franchises. It threatens to undermine the idea of collective solidarity which has historically informed public policy and the welfare state.

Secondly, solidarity's confusing humanitarian, commercial and political appeal in the neo-liberal world suggests that the State, the private sector and civil society think and act together with the poor when addressing domestic hunger and food poverty. Yet charitable food banking has been termed an 'uncritical' way of expressing solidarity (Pérez de Armiño, 2014), a term which has a substantive and concrete meaning: support for human rights, dignity, food and social justice and sharing in the universal wellbeing of the poor. One may well wish to ask whether these are the goals or functions of the corporately captured charitable food bank movement. Do they express in public policy and practice a true sense of collective solidarity with the poor?

Awkward questions

The fact is there are those from within the food bank movement itself who do not believe they are standing shoulder to shoulder with those in the breadlines. They pose awkward questions. Some, admittedly small in number, have decided to close their food banks. The busy NG7 food bank in Nottingham, UK food shut its doors at the end of 2014 in protest against 'the city council justifying welfare cuts on the grounds that desperate people can turn to food banks instead' (Owen, 2014). NG7 believes there are more effective ways to fight for food justice and food sovereignty. In Canada Operation Sharing in Woodstock, Ontario closed its food bank and set up Food for Friends which collects cash donations which funds food cards given to recipients enabling them to choose the food purchases at local food stores (Keenan, 2015). In Spain concern has been expressed about the stigma of receiving food hand outs which fails to respect confidentiality. In fact as Pérez de Armiño goes on to say in some provinces 'the catholic organization Caritas has decided not to distribute food issued from banks, opting instead to distribute money to people in need' (Pérez de Armiño, 2014, p.137). Some food banks, highly dependent on volunteer labour, go out of business as demand is always outstripping the supply of surplus food.

However it is Freedom 90 in Ontario, the union of food bank volunteers which perhaps best exposes the fiction that food banks are standing tall with the poor and hungry. Interestingly the goal is not to close food banks as of today but to make them obsolete so that its volunteers may retire before they research 90 years of age! As unpaid workers within the system Freedom 90 argues not only that charitable emergency food aid has now become permanent but that it constitutes 'a separate and segregated food system for people with low incomes' which is 'undignified and often humiliating' and 'is also inefficient, inadequate and unsustainable. The existing food distribution system – stores and markets – is inaccessible to many people simply because they lack enough money' (F90, 2017). As Freedom 90's website makes clear hunger is rooted in poverty and charity is ineffective. It is the responsibility of government to ensure basic human needs are met, encompassing a right to health and dignity and enough money to pay the rent and purchase food (ibid), in other words the right to an adequate standard of living. If food bank volunteers are thinking and acting along these lines is that not an invitation to the rest of us?

In my view Freedom 90 accepts the moral imperative to feed hungry people yet its Charter and public actions reflect an understanding of solidarity which is both critical and collective reaching beyond charity. It reaches beyond the presumed solidarity of creating secondary food markets for secondary consumers with secondary food. It is also about more than simply questioning the effectiveness and ethical dilemmas of the corporately captured and 'uncritical' charitable food aid box. It is rather about asserting the need to put the politics back into hunger, in the process demanding that government and those elected to serve the public interest, reset their moral compass.

Critical and collective solidarity: we are all rights holders

The solidarity which informs the human right to adequate food is both critical and collective. 'Critical' in that it questions the hidden functions of corporate food banking – who really is benefiting and why; and because there is a continuing and unresolved crisis of domestic hunger, the dark hole at the centre of neoliberalism, which urgently requires the attention of the State. 'Collective' in that comprehensive – 'joined-up', 'integrated' – public policies are needed successfully to address food poverty which must be grounded in human rights expressing a common humanity, that is respect for human worth and dignity. We are all rights holders.

Beyond surpluses

Given that modern day food banking has its North American and European origins in social Catholicism as in St Mary's Food Bank in Phoenix and the Paris food bank, it is useful to ask whether the idea of working in solidarity with the poor should be about more than delivering charitable food parcels and hampers to the hungry, and building bigger and better food banks.

I am not a Catholic but I believe the words of Pope John Paul II have more than a little resonance. In addressing the needs of the poor and the promotion of justice, John Paul argued that 'it is not merely a matter of "giving from one's surplus", but of helping entire peoples which are presently excluded or marginalized to enter into the sphere of economic and human development. For this to happen, it is not enough to draw on surplus goods which in fact our world abundantly produces; it requires above all a change of lifestyles, of models of production and consumption, and of the established structures of power which govern our societies' (John Paul II, 1991)

Significantly the current Pope Francis adds weight to the former pontiff's plea. In commenting about globalization in his Message for the Celebration of the World Day of Peace in 2014, Francis noted 'the many situations of inequality, poverty and injustice, are signs not only of a profound lack of fraternity, but also of the absence of a culture of solidarity. New ideologies, characterized by rampant individualism, egocentrism and materialistic consumerism, weaken social bonds, fuelling that "throw-away" mentality which leads to contempt for, and the abandonment of, the weakest and those considered "useless". In this way human coexistence increasingly tends to resemble a mere *do ut des* which is both pragmatic and selfish' (Franciscus, 2014, pp.2–3).

These are challenging words when thinking about institutionalized corporate food charity in affluent societies, and who is benefiting and why. They invite us to think about the concept of 'throw-away' food for 'thrown-away' people ('left over' food for 'left behind' people) and what it means to be on the receiving end of such largesse. At the same time Francis's observation that 'I give in order that you give' (*do ut des*) is akin to a *quid pro quo* (O'Hogan, 2017). In the food banking world it is certainly a moral and pragmatic response to persistent domestic hunger, but also contains a self-interested quality. As has been perceptively observed 'food banks provide companies with a space for their activities of "solidarity marketing" and "corporate social responsibility" with which to transmit the fact of their adopting certain attitudes and values desired by society to win consumer loyalty and improve their results' (Pérez de Armiño, 2014).

There is an expectation by the corporate food bank donor of receiving something in return, benefiting from lower food waste landfill costs, tax incentives and increased profitability. The moral safety valve of charitable giving allows individuals and our corporate food sponsors to believe we have done our bit while the structural issues of unemployment and low incomes remain untouched, and politicians look the other way. Giving and giving as a form of selective or micro-solidarity is not enough. As the anonymous on-line Tyee commenter noted 'the many acts of charitable goodwill prevent us from asking who is accountable for this morally unacceptable state of affairs' (Riches and Graves, 2007)

Importantly Pope Francis does not leave the matter there. He points to an answer when discussing 'the serious rise of relative poverty' (p.5) even though absolute poverty might be declining. From a social justice perspective he argues for effective policies to 'to promote the principle of *fraternity*, securing for people – who are equal in dignity and in fundamental rights – access to capital, services, educational resources, healthcare and technology so that every person has the opportunity to express and realize his or her life project and can develop fully as a person' (p.5). Tellingly he advocates for 'policies which can lighten an excessive imbalance between incomes' (Franciscus, 2014, p.5).

Cautiously expressed but Pope Francis' reference to 'policies' shows the way ahead. In the gilded age of neo-liberalism, solidarity with the poor requires the State to become re-engaged recognizing that we are all rights holders. In the context of first world hunger the papal message is a plea to governments to step aside from their indifference which for too many years has left the poor to the vagaries of food charity and corporate philanthropy with their bundle of conflicting interests. He is reminding the State of the need to reset its moral compass and embrace the idea of critical and collective solidarity through progressive public policy informed by human rights. It is about moving beyond surpluses and changing the conversation from one of charity (corporately captured) to a re-imagined idea of collective solidarity informed by the right to food.

Moral obligations and political commitments

The point is surely that fully to address hunger and poverty requires a 'social, political and moral commitment' by the state. This appeal was made by Louise Arbour, a former member of Canada's Supreme Court and the UN High Commissioner for Human Rights (2004–2008) in her seminal 2005 LaFontaine-Baldwin Lecture, entitled *'Freedom from Want – from Charity to Entitlement'*. Part of her speech was defending the progress made in Canada in the post WWII era regarding the role played by the federal state in establishing universal health care, social insurance for the unemployed and advancing social services which she notes were 'special types of goods, not commodities which could be made contingent upon the ability to pay' (Arbour, 2005 p.4).

Arbour, however, was not making the case that Canada's social history from 'the early to mid 20th Century was a bygone era of cooperation and solidarity', but rather that 'social transformations demand political commitment', individuals working together 'for their ideals, who hold politicians to account, and who defend principles wherever they are, from the kitchen table to the international stage' (Arbour, 2005, p.3), and particularly when defending universal rights and freedoms. Universal health care she argued is a cherished national institution (doubtless the case in other societies) which 'we see as a cornerstone of Canadian values, a way of honouring our fundamental commitment to each other' (ibid, p.4). This is a reasoned argument for collective solidarity which informed the origins of the post WWII welfare state. In the immediate post war period it captured the idea of a home fit for everyone, the idea that we are all in this together.

More specifically, Louise Arbour argued, that freedom from want requires building the necessary public institutions and social policies which are 'a matter of obligation at law owing to a duty which goes to the core of the protection and promotion of human dignity' (ibid, p.4). Charity, while always with us, was an inadequate long-term response to the causes of hunger and poverty. At the time of the global food crisis in 2008 she reminded the UN Human Rights Council that at its core and in its punitive effects, it was essentially a matter of a lack of access to food, and that such access is a right protected under international law.

While her message was principally directed at the poor world, it was one with equal ramifications for rich world food bank nations which had been ignoring the human right to adequate food by leaving domestic hunger to the happenstance of charity. Arbour was calling for co-operation and solidarity which found expression in the UN Charter, the *ICESCR* and the *MDGs* (Arbour, 2008). The UK is a prime example of such state neglect. As Elizabeth Dowler has written 'the official government response to the growth of food banks has been to deny they are part of the welfare system' … moreover 'there is no recognition by the central state of the problems associated with, nor acceptance of responsibility for the growing number of households unable to survive economically, and who are using charitable and emergency food services' (Dowler, 2014, p.169).

This was stated more bluntly in a headline for a 2014 Guardian article by Zoe Williams about the current state of child hunger in the UK and the violation of human rights. It asks 'as children starve, where's the state?' Human rights, she argues, need to be back on the table, 'some framework of universal entitlement, established by the principles of justice, solidarity and humanity would actually be helpful' (Williams, 2014). This is a timely reminder that we are all rights holders who should be demanding that the State demonstrates a moral commitment informed by a sense of collective solidarity. Given that each of the OECD food bank nations, with the exception of the United States, has ratified the right to food, such pleas for governments to reset their moral compass requires priority attention. Williams is also reminding us that the right to food is a matter of international law.

Why the right to food in international law matters

Objectively speaking the right to food matters because it has been established in international convention and law for more than fifty years, and longer as first articulated in the UDHR (1948). The right to food is not mere rhetoric. It has concrete legal definition, establishes legitimate claims upon governments and has been ratified by 164 UN member states. It is justiciable. It confers entitlement. The right to food enjoys broad universal support and is a bedrock component of economic, social and cultural rights and the right to an adequate standard of living. As a fundamental human right it recognizes the inherent dignity and worth of all human beings. The right to food matters because it is an expression as well as a means of bringing about collective solidarity. It moves beyond charity and the corporate donation of 'left over' food and engages issue of moral, legal and political commitments and obligations.

Yet, in the context of domestic hunger and food poverty in today's neo-liberal age, the principal reason why the right to food matters is because the state is missing in action, and hardly to be found. Its neglect, or luke warm cheering on from the sidelines, has spawned the spread and institutionalization of corporate food charity reliant on surplus food to feed bodies and minds. While seemingly enjoying public legitimacy the use of surplus food by charitable food banking has proved unequal to the task of addressing domestic hunger and food insecurity, to say nothing of reducing poverty. The corporate capture of charitable food banking has become part of the problem. Even leaving aside the moral legitimacy of using wasted food to feed hungry people, the prevalence of food poverty is beyond the capacity of philanthropic endeavour. The right to food matters because food matters in all aspects of an individual's life from cradle to grave and that food security is a primary obligation of the state.

Basic human need and fundamental human right

It must first be emphasized that food is a *basic* human need and is universally recognized as fundamental human right based on the 'inherent worth' of the individual. Basic in the absolute sense, like water and air, of being essential to human growth and development, to individual health and social well-being. Food lies at the centre of family, community and cultural life. To be denied access to food results not only in the loss of nutritional well-being but of one's personal, social and cultural identity. Its lack robs people of the means to earn their livelihoods, to purchase the food of their choice and feed themselves with dignity. Being deprived of food excludes people from participating in normal and customary ways in society, and most profoundly is an affront to human dignity. In words attributed to Eleanor Roosevelt 'a "right" is not something that somebody gives you; it is something that nobody can take away'. The right to food is an expression of collective solidarity.

What the right to food is

Interestingly, Geraldine van Bueren, QC, professor of human rights at Queen Mary's College, London, notes that in 1216, eight hundred years ago, 'English law first recognized a right to food', yet England has forgotten its legal history. She points this out whilst noting that in September 2013 'over 350,000 people received three day's emergency food from the Trussell Trust food banks, triple the numbers helped in the same period' the previous year (van Beuren, 2013). This protection of the right to food, she writes, is to be found in *The Charter of the Forest*, which 'was regarded as a sister Charter to the *Magna Carta*' (1215) and comprised rights such as 'the right to honey, grazing rights and rights to firewood, (which) constituted the essentials of medieval life, but translate in the 21st century to the right to adequate nutrition' (van Beuren, 2013).

Undeniably for the last seventy years the human right to adequate food has been written into international law dating from the 1948 *Universal Declaration of Human Rights* and in 1966 the signing of International Covenant on Economic, Social and Cultural Rights (Box 8.1). It is grounded in internationally agreed and ratified standards, obligations and responsibilities informed by human rights principles: universality, human dignity, autonomy, participation, accountability, empowerment, non-discrimination, transparency and the rule of law, which recognize the indivisibility of all human rights.

BOX 8.1 OPERATIONALIZING FREEDOM FROM WANT

- **UDHR Article 25:**
  - Everyone has the right to a standard of living adequate for the health and well-being of himself and of his family, including food, clothing and housing and medical care and necessary social services

- **ICESCR Article 11:**
  - 11.1 The State Parties recognize the right of everyone to an adequate standard of living for himself and his family, including adequate food,

> clothing and housing, and to the continuous improvement of living conditions;
> – The State Parties recognize the fundamental right of everyone to be free from hunger

As stated by UN *General Comment 12*, authored by leading food and human rights experts commissioned to clarify the meaning of the right to food in international law, the right to food is 'realized when every man, woman and child, alone or in community ... have physical and economic access at all times to adequate food or the means for its procurement'(CESCR, 1999). It is about the right to feed oneself and one's family with dignity and choice with adequate financial resources, that is to produce or acquire food in normal and customary ways, in the grocery store, the supermarket or the farmer's market.

It comprises the right of every one to an adequate standard of living, including adequate food; and the fundamental right of everyone to be free from hunger (Ekwall, 2008). It is 'precisely laid out' in the *International Covenant on Economic, Social and Cultural Rights (Article 11, ICESCR, 1966)* as a key element of the right to an adequate standard of living (Ziegler et al, 2011) with cross cutting rights as in international legislation comprising the *Convention on the Elimination of all Forms of Discrimination Against Women (CEDAW, 1979)* the *Convention on the Rights of the Child (CRC, 1989)*, the *Convention on the Rights of Persons with Disabilities (CRPD, 2006)* and the *Optional Protocol to the International Covenant on Economic, Social and Cultural Rights (OPESCR, 2008)*. Notably none of these conventions or their protocols have been ratified by the USA.

Ratification of the *ICESCR* by all OECD governments, excepting the USA, requires them to act in compliance with their obligations under international law to ensure the progressive realization of the right to food and the achievement of food security for all. It is the primary role of government to ensure all people have the capacity to feed themselves in dignity: enabling all to produce and acquire food through normal and customary ways, that is with enough cash in their pockets to purchase the food of their choice. Yet food charity slowly undermines the idea that domestic hunger is a collective responsibility and a lawful right.

A legal right: justiciability

The right to food is not only aspirational but is also a legal right, a justiciable claim actionable through the courts. A right is not a right unless it can be claimed. It is an entitlement. It is not about charity which cannot be claimed. It is the right of everyone to have a judicial or other effective remedy when their rights have been violated (see UDHR, Article 8). For example when individuals are unable to feed themselves or their children as a result of inadequate welfare benefits or the imposition of sanctions they must be able to appeal such decisions either through social security tribunals or the courts.

ICESCR: constitutional entrenchment

However, for such claims to be brought before the courts with any reasonable chance of success, to say nothing of the length of time such an appeal may take, the right to food is required to be entrenched within the national constitutions of individual ICESCR state parties.

Interestingly as Table 8.1 indicates only Mexico as an OECD member state has explicitly incorporated the right to food into its national constitution. The majority of the OECD food bank nations either implicitly provide such recognition (e.g., through a standard which does not fall below a minimum subsistence level) or by pledging to do so in the form of directive principles of state policy but which are 'understood to be non-directly enforceable by a court'. Some states may also offer a higher status to international obligations over national legislation (see FAO, 2017b and Table 8.1, below).

However, as the rise of the OECD food bank nations reveals whether or not the right to food is constitutionally entrenched, explicitly or implicitly recognized, subject to directive principles or granted the higher status of international law, or as in the case of Australia, Canada, Chile, New Zealand, the UK and the USA, it seems to make little difference. Domestic hunger grows, corporate food charity marches on and governments continue to violate the right food and freedom from hunger. Does the law matter, one might well ask. Do the courts have a role? In practice how justiciable are economic, social and cultural rights within affluent nation states and indeed in the poor world?

OPESCR: international recourse mechanism

In fact, the question of justiciability has received the attention of the OHCHR and the CESCR for a quarter of a century leading in 2008 to the adoption of the *Optional Protocol to the International Covenant on Economic, Social and Cultural Rights* (*OPESCR*) by the UN General Assembly. The global NGO community including

TABLE 8.1 Promises, promises: constitutional recognition of the right to food in OECD food bank nations, 2017

Recognition	State parties
Explicit protection	Mexico
Implicit protection	Belgium, Czech R, Finland, Greece, Iceland, Italy, Japan, Netherlands, Portugal, Slovakia, Spain, Switzerland, Turkey
Directive principles	Germany, Ireland, Latvia, Norway, Poland, Korea, Sweden
National status of international obligations	Austria, Estonia, France, Hungary, Luxembourg, Slovenia
No report	Australia, Chile, Canada, New Zealand, UK, (USA)

Source: derived from FAO database: FAO, 2017b

TABLE 8.2 Signatories and parties to the OPESCR by OECD food bank nations, 2017

Status	State parties
Signatories	Ireland, Netherlands, Slovenia
Parties/Ratification	Belgium, Finland, France, Italy, Luxembourg, Portugal, Slovakia, Spain

Source: UN Treaty Collection, 18 June, 2017
https://treaties.un.org/pages/ViewDetails.aspx?src=TREATY&mtdsg_no=IV-3-a&chapter=4&clang=_

such organizations as Amnesty International (AI), FIAN International and the Social Rights Advocacy Centre in Canada have played a significant role in the creation of the *OPESCR* which entered into force in 2103. Today it has 45 signatories and 22 state parties have ratified the protocol (*OPESCR*, 2017). Of these as Table 8.2 shows it has been signed (3) and ratified (8) by eleven EU countries, a little under a third of all OECD food bank nations (see Table 8.2, above).

Given the lack of enforcement by, or weakness of the courts to legally enforce the right to food or housing or the adequacy of benefits within particular nation states, the *OPESCR*'s primary purpose is to create a complaints mechanism and legal remedy at the international level – the UN CESCR – to which individuals might bring their grievances. As the Worldwide Movement for Human Rights expressed it 'if national courts fail to protect against violations of their economic, social and cultural rights, people will be able to raise their case with the UN CESCR' (Fidh, 2013). Individuals may only bring their case forward if they have exhausted all domestic remedies. The process rules out anonymous complaints and matters which had occurred prior to ratification.

The aim of *OPESCR* is also to strengthen the development of effective national remedies and clarify State obligations and access to justice at the national level (see Fidh, 2013). The NGO Coalition for the *OPESCR* also commented that after 40 years economic, social and cultural rights had now achieved parity with the equivalent *Optional Protocol to the International Covenant on Civil and Political Rights*, thereby giving true meaning to the indivisibility of and interrelatedness of all human rights (NGO-ESCR, 2013). In other words justiciability had been put on an equal footing. As such it expresses an idea of collective solidarity.

What the right to food is not

Importantly, understanding the right to food as a bundle of moral, legal and political commitments and obligations under international law tells us what the right to food is not. It certainly does not provide a legalistic quick fix for domestic hunger and food insecurity, rather its implementation requires policies and actions in many fields, not only the legal field. Indeed full implementation should permeate all sectors of government and civil society.

Nor is the right to food about government doing everything for everyone. As Mary Robinson, the former President of Ireland (1990–97) and UN High

Commissioner for Human Rights (1997–2002) has written 'the right to food is not about giving away free food to every body' (Robinson, 2004). It is not about the right to be fed and should not be equated with food aid which should be an emergency measure with a limited time frame (RFU-UNFAO, 2008). In other words the right to food is not about charity.

State accountability: resetting the moral compass

As Louise Arbour has argued 'there will always be a place for charity but charitable responses are not an effective, principled or sustainable substitute for enforceable human rights guarantees' (Arbour, 2005). This much is clear however powerful the transnational corporate food champions of charitable food banking may be. Indeed, the right to food poses difficult questions about the role played by the parallel food charity economy in stigmatizing the poor whilst undermining the welfare state and publicly funded social safety nets. It draws attention to the rise of the food bank nation, its corporate take over and supervision of the new charitable food safety net: the shift from cash transfers and income security to food transfers and long-term emergency food aid.

Charity is certainly about the moral imperative to feed hungry people. However in light of this unacknowledged and undemocratic transfer of responsibility for social protection and welfare policies to Big Philanthropy and Big Food, the right to food demands answers from governments. These principally concern the accountability and obligations of the State for safeguarding human rights and ensuring that all people have the income necessary to meet their basic needs including feeding themselves and their families with choice and dignity.

When charitable food aid has become a key institutional component of a country's food security policies (by design in the USA and negligence in other OECD nations), challenging as it is, the right to food in international law provides an opportunity to reset the moral compass of the State for addressing domestic hunger and achieving food security for all. After all, government is meant to be in charge.

Reflections

From a perspective of critical and collective solidarity informing the right to food, this chapter has argued that the conventional and rhetorical idea of solidarity as the moral driver of corporate food charity as an effective first responder to widespread and long-term domestic hunger needs urgent reconsideration There is a need to move 'beyond surpluses' to embrace the moral, legal and political obligations of the State as established in international law. The right to food as clarified within the *International Covenant on Economic, Social and Cultural Rights* together with its *Optional Protocol* provides a framework of international and national law obligating austerity focused governments of affluent nations to be publicly accountable.

Whether or not governments, rich and poor alike, have constitutionally entrenched the right to food into domestic law, when they ratify conventions such as such as the *ICESCR* and *CRC* they are required to act in domestic compliance with their obligations under international law to ensure food security in the own countries. The fact is we are all rights holders and there is a need for governments to reset their moral compass. It is a matter of public accountability and of using international law to hold governments to account.

9

PUBLIC ACCOUNTABILITY AND THE RIGHT TO FOOD

International monitoring to the rescue

Looking back over 35 years it is clear that changing the conversation about domestic hunger in the rich world from food charity to one of collective solidarity, public accountability and the right to food is not a task for the faint hearted, particularly when seeking the attention of the State. That much is confirmed by many years of neoliberalism, austerity and the shrinking welfare state. As OECD food bank nations have multiplied, regrettably the shameful State, as Stefan Selke has described it (Selke, 2015), continues to look the other way, indifferent to the plight of hungry people eating surplus food.

Yet the right to food discussion remains critical if public debate is to be informed and negligent and unaccountable governments are to be held to account. Surely the goal must be that affluent governments accept their moral obligations to realize the right to food for all in the context of national food security, paying particular attention to the needs of vulnerable populations unable to feed themselves. In other words, making sure no one is going hungry and they have access to affordable and nutritious food.

The good news is that if governments wish to reset their moral compass, there is plenty of expert advice and support on hand. International food policy and human rights agencies such as the UN Food and Agricultural Organization, the Human Rights Council and Office of the High Commission for Human Rights, the Special Rapporteur on the Right to Food, and the periodic review processes of the HRC and the Committee on Social and Economic Rights offer considerable possibilities as non–State mechanisms and tools for monitoring, dialogue and advice.

Food, as a matter of human rights

UN human rights conventions and protocols such as the *ICESCR* and *OPESCR* safeguard the right to food,. They provide a framework of rights-based international law requiring the compliance of State Parties in taking action to eliminate domestic hunger. Governments in ratifying these instruments will acknowledge they are rooted in shared human values, planning, the rule of law and democratic accountability. Yet on the matter of the State's obligation to ensure the progressive realization of the right to food enthusiasm wanes, fearful that its national constitutional entrenchment will undermine sovereignty and domestic policy-making.

It is therefore instructive to note the comments of Jacques Dioff, the former FAO Director-General, regarding the advice adopted in the FAO's *Voluntary Guidelines* given their intent to provide practical guidance to member states regarding rights-based approaches to poverty reduction and the achievement of food security for all. *The Right to Food Guidelines*, he wrote, 'cover the full range of actions to be considered by governments at the national level in order to build an enabling environment for people to feed themselves in dignity and to establish appropriate safety nets for those unable to do so'. He emphasized the social and human dimensions of development, 'putting *entitlements* (author's emphasis) of people more firmly at the centre of development' (FAO-VGs, 2005).

In the struggle against global hunger this was undoubtedly a message to the poorer countries of the world. However, more than a decade after he wrote these words, and in light of widening inequality, the First World's broken social safety nets and the failure of food charity to solve widespread domestic hunger, the notion of entitlement – the right to food – resonates even more strongly in today's high income states. Specifically with respect to food access Dioff reminds us that ESCR law calls to the attention of all member states, including the OECD food bank nations, their obligations for compliance with international law, and the primary role of government as the accountable State.

The state as 'primary duty bearer'

Whilst the history of human rights proclaims we are all '*rights holders*', it is government which is the '*primary duty bearer*' accountable under international law for the eradication of hunger and poverty. Certainly a multi-stakeholder approach inclusive of civil society and the private sector is recognized and is essential (FAO-VGs, 2005). Yet it remains the primary obligation of the State and public policy to ensure food security for all inclusive of the most vulnerable populations, whether citizens or non-citizens inclusive of refugees and undocumented migrants.

Ratification of the *ICESCR* is recognition by State Parties of their moral and legal obligations as the *primary duty bearer* in their respective countries to act in compliance with international law to *respect, protect and fulfil* the progressive realization of the right to adequate food in the context of national food security (UNESC, 1999; -VGs, 2005; Arbour, 2008). Most importantly for governments and public advocacy, 'rights rhetoric provides a mechanism for reanalysing and renaming "problems" as "violations", and, as such, something that need not and should not be tolerated' (Jochnick, 1999, p.60 quoted in Dowler and O'Connor, 2012, p.46).

Respect, protect, fulfil

General Comment 12 makes clear the specific obligations of government regarding its role in monitoring and ensuring domestic compliance to 'respect, protect, fulfil' the right to food (Ziegler et al, 2011; see Riches and Silvasti, 2014):

- *Respect* the right to food means 'that Government should not arbitrarily take away the peoples right to food or make it difficult for them to gain access to food'. In other words welfare states must not restrict people's access to food by enforcing or tolerating sub-poverty minimum wages; zero-hour contracts and inadequate welfare benefits; sanctions, administrative delays and denial of assistance and social spending cuts.
- *Protect* the right to food requires Government to 'pass and enforce laws to prevent powerful people or organizations from violating the right to food'. In other words regulating 'non-State actors', including corporations or individuals who may threaten other peoples' right to food, for example, protecting the food sovereignty of Aboriginal populations, agricultural land, and ensuring food safety.
- *Fulfil* the right to food obliges Government to 'take positive actions to identify vulnerable groups and to implement policies to ensure their access to adequate food by their ability to feed themselves'. This not only includes stimulating employment and protecting workers' rights but taking the necessary legislative, budgetary and administrative steps to develop and implement national policies ensuring a living wage, adequate benefits, social housing, universal child care, including progressive taxation and acting as the provider of last resort in terms of social protection (Riches and Silvasti, 2014).

Broadly speaking as Ziegler et al have noted this means that governments must not take actions that result in increasing levels of hunger, food insecurity and malnutrition. They 'must protect people from the actions of others that might violate the right to food; and to the maximum of available resources, invest in eradicating hunger' (2011). In short, the right to food demands that domestic hunger is addressed as a deeply concerning moral, legal and political matter within affluent societies; and as a primary obligation of the state. The State holds the collective responsibility and is publicly accountable for guaranteeing that basic needs are met, and that the right to food is fully realized.

The indifferent state: looking the other way

Obviously, against a 30-year backdrop of an increasingly dysfunctional industrial food system, broken social safety nets and regressive income distribution it is one thing to declare the negligent State as the 'primary duty bearer' for addressing domestic hunger (and food waste), and quite another for it to move from its shameful indifference to public accountability. It faces a challenging mandate even if it is morally and ideologically so inclined. In searching for effective remedies and courses of action there are many obstacles in the way.

Firstly, as we have noted, the call to entrench the right to food in national constitutions (see Ch.8) while of principal significance remains a most unlikely occurrence amongst indifferent OECD food bank nations. Only Mexico has stepped forward. The majority of countries claim either explicitly or implicitly to guarantee this right either through directives or broader interpretations of human rights, or even not at all. Inevitably there are many loopholes as indicated by their dependence upon parallel systems of corporate food charity and food safety nets.

Constitutional entrenchment would strengthen the development of national framework legislation for food policy. As the former UN Special Rapporteur for the Right to Food has noted 'the Committee on Economic, Social and Cultural Rights has insisted on the need for States to work towards adopting "a national strategy to ensure food and nutrition security for all, based on human rights principles that define the objectives, and the formulation of policies and corresponding benchmarks"' (De Schutter, 2008). He remarks this includes connecting the dots between different ministries and levels of government focused on coordinated national action plans; improving accountability and ensuring the participation of the food insecure in the progressive implementation of the right to food.

In terms of preventing and responding to the daily experiences of domestic hunger and food insecurity such courses of action would include progressive income distribution, integrated nutrition, public health and social policy including such measures as a living wage, adequate social security, affordable housing and child care. Such 'joined-up' policy-making seems a long way off. Stating it bluntly, for the indifferent State food is a political football too easily kicked from one ministry silo to another, frequently missing the goal by being punted out of touch as the unconcerned referee just looks away.

The incorporation of the right to food into national law would also assure its justiciability as an actionable legal concept offering judicial or quasi-judicial remedies through the courts, human rights tribunals or welfare appeal systems. In this way the State could be held to account by an independent judiciary for its compliance with international law, for example protecting vulnerable individuals whose access to food may have been be violated by inadequate income security benefits, underfunded social safety nets and the imposition of punitive welfare policies. Yet the indifferent State looks on. There is little appetite for entrenching justiciability and achieving a rights–based system of social protection.

Secondly, even in the EU with its commitment to solidarity, the right to food is missing in action. Not only does the section on Solidarity in the *Charter on Fundamental Rights* (CFREU, 2000) lack any mention of the right to food, but as noted by Oliver De Schutter 'it is weak, in general, on social rights' (ODS, email communication 9.3.2016). Likewise its absence is noticeable in the EU Commission's recently proposed *European Pillar of Social Rights* (EPSR, 2017) designed to deliver new and more effective rights for citizens. Amongst a bundle of social rights protecting vulnerable populations, the right of children to be free from poverty and their right to adequate minimum income benefits is recognized. Yet the commitment to the right to access essential services of good quality, comprising 'water, sanitation, energy, transport, financial services and digital communications' contains no mention of food. Solidarity cannot be counted upon.

The exclusion of the right to food from EU law should be a wake up call to the EU Commission and MEPs in light of its recognition by the United Nations and its full embodiment and status in international human rights law. Perhaps it is because food is so available and every day for EU politicians and those drafting and embedding such principles as solidarity and social rights into EU documents that food just gets taken for granted. Neglected as a basic human need and fundamental human right, hunger is not a matter for the State's concern. Such indifference is a necessary reminder that solidarity, if it is to have teeth, needs to be both critical and collective.

Thirdly, the effectiveness and achievements of public policy in reducing and eliminating domestic hunger and food insecurity should be informed by evidence-based research. Its lack in the majority of OECD countries is a significant obstacle. As empirical studies have shown food bank usage data is an unreliable measure for determining the prevalence of national food insecurity, and therefore for developing public policy which works. It is necessary to establish officially recognized indicators of food insecurity or food poverty with verifiable benchmarks, goals, targets and timeframes for measuring and monitoring the reduction of food poverty. Still, even in North America which has valid and reliable data concerning food insecurity, neoliberal minded politicians, indifferent or unwilling to address the failure of charitable food banking, continue to look the other way. Political will which is in short supply, is needed to first grasp the issue of domestic hunger, research its prevalence and determine what needs to be done.

Such obstacles present considerable challenges for any nation state whether poor or rich with a mind to entrench and implement the right to food. However there is a body of internationally agreed law which recognizes economic, social and cultural rights including the right to food. The *ICESCR* (1976) has been ratified by 164 UN state parties including all but one – the USA – of the OECD food bank nation states. The *ICESCR, CEDAW*, the CRC and the UN Declaration of the Rights of Indigenous Peoples (2007) keep the right to food firmly on the inter-national development table. They create the public and political space necessary for informed dialogue and democratic debate about domestic hunger and food insecurity and how it should be addressed from a human rights perspective. Understandably the space is largely occupied by nation states in the Global South but it is one which could be of immense benefit to the wealthy throwaway societies of the North.

It is an invitation to discuss the meaning and practical implications of critical and collective solidarity by providing the opportunity to engage discussion about the right to food and the moral, legal and political obligations of the state to address widespread domestic hunger and achieve food security for all. It is about framework legislation and national food policy; justiciability and effective remedies; the effectiveness and moral dilemmas of using surplus food to feed hungry people; the benefits of 'joined-up' food, public health and social policy; publicly funded social protection programmes for vulnerable populations and much more.

This of course requires a huge dose of political will but it is important to recognize the role played by the United Nations in mainstreaming human rights in the struggle against hunger and poverty in poor and rich countries alike. In particular the UN's right to food analysis and focus on national framework legislation and integrated food policy suggest courses of action for moving beyond dysfunctional food systems and broken social safety nets and that the right to food has legs.

Mainstreaming the right to food: international mechanisms

In addressing issues of global hunger and food insecurity the mandates of the UN Human Rights Council and the FAO play important roles in transforming the right to food from being a matter of international law (*ICESCR*) to one of national policy and practice. It is fair to say that the FAO through its right to food approach to development work in the Global South has contributed to raising awareness about a rights-based analysis of domestic hunger and food insecurity in the North, ironically perhaps, a reverse or reciprocal form of food aid.

FAO Right to Food Team

As a response to the 1996 and 2002 World Food Summits, an *ad hoc* Right to Food Unit was established in 2003 as a special unit attached to the FAO's Economic and Social Department in Rome. Its task was to support the negotiation process developing the Right to Food Guidelines. After the adoption of the *Voluntary Guidelines* in 2004 the *ad hoc* became permanent and the Right to Food Unit was established (Frank Mischler, personal communication 23.6.17). Later the Unit became the Right to Food Team and is currently housed in the Social Policies and Rural Institutions Division (ESP) of the Department of Social and Economic Affairs where it coordinates its work with teams using rights-based and capacity-building approaches to strengthen 'rural institutions, services, social protection, gender equality, decent rural employment and the right to food' (FAO-ESP, 2017).

It was the pioneering work of the Right to Food Unit that gave a significant global boost to rights-based approaches to agricultural development work, access to food and poverty reduction. Today this is reflected in the FAO's use of the right to food as a normative and analytic framework which directs its activities in eliminating hunger and advancing food security in the poor world. Of particular interest while the *Right to Food Guidelines* are voluntary they do, however mildly, question the prevailing neo-liberal assumption that economic growth and ever-increasing agricultural production is the solution to global hunger. They point to unaddressed questions of the just distribution of food and income, gender inequality, unemployment, education and skill development and empowerment which demand attention (FAO-VGs, 2005).

Today the Right to Food Team works with regional FAO field offices and uses the guiding framework of the Panther principles (Box 9.1) in strengthening policy-making processes which guide programme formulation in their development activities.

BOX 9.1 RIGHTS-BASED PANTHER PRINCIPLES, FAO RIGHT TO FOOD TEAM, 2017

Principles	Participation, Accountability, Non-discrimination, Transparency, Human Dignity, Empowerment, Rule of Law
Programme	Implementation, monitoring and evaluation
Formulation	Strengthening the notion of rights and obligations
Process	Building the capacity of rights holders and duty bearers. Establishing accountability mechanisms

---

The Right to Food Team also produces a range of educational tools – publications, handbooks, methodological tool boxes, videos, study and facts sheets to advance its work (see FAO, IFAD and WFP FAO-RTFT, 2017). Perusing its website, one cannot help wonder whether the Panther principles, or human rights based approaches (HRBA), and its rich source of online materials should be put to good use in the world's wealthy OECD food bank nations.

In my discussion with Juan Garcia Cebolla, the Right to Food Team leader, he told me the right to food approach is universally valid with lessons for developed countries. It ensures coherence in national policy-making, guards against discrimination; it requires the use of evidence-based research – measuring food insecurity – to inform policy-making and its use of monitoring compliance and violations would be equally applicable (JGC, personal communication 21.5.17).

Human Rights Council – Office of the High Commissioner for Human Rights

The UN's Human Rights Council (HRC) and its secretariat the OHCHR is the principal UN office mandated to promote and protect human rights throughout the world. Its focus is 'mainstreaming human rights' through 'standard setting, monitoring and implementation on the ground' (OHCHR, 2017b). Given that the primary responsibility for protecting human rights lies with governments – the State as *'primary duty bearer'* – the OHCHR's role is advisory and educational. Yet at the same time its monitoring functions are designed to hold governments to account acknowledging progress when made but highlighting matters of concern and issues of human rights violations when necessary.

The OHCHR's mandate supports the work of the independent UN special rapporteurs including the position of the independently appointed Special

Rapporteur on the Right Food and conducting on site country missions in the developing world. Of note, however, since 2012 it has led country missions to Canada, Mexico and Poland, all OECD member states. The OHCHR also assists the ten UN treaty bodies and their national-level monitoring of compliance with the core international human rights treaties. This involves supporting the Committee on Economic, Social and Cultural Rights (CESCR, 2017) and its team of 18 independent experts who monitor State Parties' compliance with the *ICESCR* including the right to food; as well as the complaints mechanism made available through the *OP-ESCR*. In a separate process the OHCHR is additionally responsible for the Universal Periodic Review process established in 2006 to monitor the complete human rights records of member states.

Special Rapporteur on the Right to Food: in the rich world

As independent experts appointed by the HRC the Special Rapporteurs on the Right to Food (Box 9.2) are charged with examining and reporting back on the progress which countries are making on the progressive realization of the right to food and achieving food security. They also issue reports on themes from a human rights perspective.

BOX 9.2 UN SPECIAL RAPPORTEURS ON THE RIGHT TO FOOD, 2000–2017

Jean Ziegler (Switzerland)   2000–2008
Olivier de Schutter (Belgium)   2008–2014
Hilal Elver (Turkey)   2014–present

Source: http://www.ohchr.org/EN/Issues/Food/Pages/FoodIndex.aspx

---

An important function of the Special Rapporteur's role is monitoring 'the situation of the right to food throughout the world', identifying general trends and conducting country missions; communicating with States and other concerned non-state actors about violations of the right to food, presenting reports to the Human Rights Council and the General Assembly; and promoting the right to food at seminars, conferences and expert meetings (see OHCHR-SRRF, 2017). Principally the Special Rapporteur works in the Global South.

Monitoring, holding governments to account and raising public awareness about 'the transformative power of the right to food' as the UN food envoy Olivier De Schutter once expressed it (De Schutter, 2017), are the key responsibilities of the Mandate providing opportunities through civil society and democratic debate to challenge existing food, economic and social inequalities so as to change the conversation to the moral, legal and political obligations of the state. Reassuringly from the perspective of widespread food poverty in OECD member states since 2012 the Special Rapporteur has begun to include country missions to high and upper middle income countries.

Country missions

A key component of the Special Rapporteur's mandate is conducting country missions at the invitation of governments. Their aim as expressed on Oliver De Schutter's website 'is to assess the efforts of States in the progressive realization of the right food, to report on its findings and to propose, in a spirit of cooperation and assistance, recommendations to improve situations indentified as matters of concern' (De Schutter, 2017). They are also intended to spread knowledge of best practices and stimulate debate about the right to food within and between 'government, parliaments, and civil society organizations, including farmers' organizations' (ibid). Country reports are provided to governments, made publicly available and presented to the UN Human Rights Council.

During his six-year tenure De Schutter made 13 such official visits, two of which were to OECD member states: Mexico in 2011 and Canada in 2012, the first high income food bank nation to host such an official visit. This was a result of Canada, to its credit, honouring its commitment to its standing invitation to the special procedures of the Human Rights Council (ODeS, email communication, 27.6.17).

A brief telling of the mission to Canada illustrates the process involved and holds lessons for other high income countries in which widespread food insecurity and charitable food banking has become embedded. It is a story of the challenges confronting the mainstreaming of the right to food at the national level and the opportunities it presents for influencing public policy.

Mission to Canada

Undoubtedly the mission to Canada met the high standards expected of the UN food envoy's inquiry. The final report demonstrates the thoroughness of the consultative process, the quality of the research and the significance of its recommendations. The Special Rapporteur met with senior federal officials from the ministries of Aboriginal and Northern Affairs, Agriculture and Agri-Food Canada, Fisheries and Oceans, Health Canada, Human Resources and Skills Development, Justice Canada and the Canadian International Development Agency. Outside of Ottawa he travelled across the country consulting broadly with provincial and municipal authorities and the Nunavut Department of Health and Social Services. De Schutter also had the opportunity to meet with Aboriginal and First Nations communities in Quebec, Ontario, Manitoba and Alberta (see OHCHR, 2013a).

From a right to food perspective the report (OHCHR, 2013a) inquired into the prevalence of food insecurity in a country as wealthy as Canada; explored the possibility of national food policy framework legislation; considered issues of food availability (agricultural policies), access and adequacy (income security, social protection and public health) and the long time unresolved issues of food sovereignty confronting Aboriginal and First Nations peoples. These include the crisis of food and nutrition insecurity affecting Inuit peoples in Canada's North, exorbitant and out of reach food prices, the lack of access to traditional foods to say nothing of historical land rights' issues.

Drawing on official national surveys the Special Rapporteur confirmed widespread and growing food insecurity across the country, and particularly its deplorable incidence among First Nations Inuit peoples but noted its absence for the Métis population. In commenting on the 900,000 people dependent on charitable food banks De Schutter noted this 'reliance on food banks was symptomatic of a broken social protection system' which served as 'a moral safety valve' for the State.

Clear-sightedly and persuasively he argued that Canada had survived better than most from the 2007–2008 global financial crisis, while questioning the government's lack of compliance in implementing the right to food. Diplomatically while pulling no punches the Special Rapporteur noted that 'Canada's record on civil and political rights has been impressive' while 'its protection of economic and social rights, including the right to food, has been less exemplary' (OHCHR, 2013a).

The report and its recommendations were well received by national civil society organizations such as Food Secure Canada and Canada Without Poverty as well as provincial and local human rights and advocacy groups. They were encouraged to have expert international backing for their analyses, reasoned arguments and advocacy over the years as to why agricultural, food, nutrition, public health and social policies adversely affecting society at large and especially vulnerable populations had to change. The report certainly presented an opportunity for public and political dialogue yet one at that time which faced considerable challenge.

The right wing Conservative Government of Stephen Harper was definitely not on side. It was hostile and abruptly dismissive both of the mission itself and of the report. Whilst the leaders of the Liberal and NDP opposition parties met with him, no Cabinet members did. Some ministers called De Schutter 'ill-informed', 'patronizing' and 'completely ridiculous'. He was advised him 'to concentrate on places like Bangladesh, China and Ethiopia where people are feeling real hunger' (Hunter, 2012). This comment simply displayed their ignorance as Jean Ziegler, the first Special Rapporteur, had already conducted missions in these countries. Even though the government of Canada had extended its invitation, its response was no surprise. As De Schutter noted, he 'arrived in the country at a politically tense time [and] was sort of caught in the crossfire – used conveniently by the opposition, and dismissed by the government, which, at the time, was very critical of multilateralism' (ODeS, email communication, 27.6.2017).

Iain Hunter, a *Victoria Times Colonist* columnist was right when 'De Schutter told us he heard from the vulnerable among us and what we already know. Canada is a rich country with a dirty secret. A lot of people – including those with the constitutional responsibility to do something about it – resent being reminded of this. Poverty lurks in the Great Canadian Gut like a tapeworm, sucking nourishment from those who need it most. De Schutter is right. It is shocking. If our governors aren't ashamed, a lot of Canadians are' (Hunter, 2012).

Perhaps such a response by government is likely to dissuade other OECD member states from issuing invitations to the Special Rapporteur on the Right to Food. They would not be worth the value of the UN calling card. This would be a mistake. Governments come and go but the official reports of the country missions remain not only as advice to future administrations but likewise to civil society, incrementally enriching research, raising bureaucratic and public awareness and assisting in changing the conversation about food policy and domestic hunger to the moral, legal and political obligations of the State. Indeed as De Schutter noted at the end of his mission 'the right to food is about politics. It's not about technicalities. It's a matter of political will. I think these comments are symptomatic of the very problem that it is my duty to address' (CBC, 2012).

Whilst food insecurity in Canada today is still estimated to affect 4 million people, it is important to note that Justin Trudeau's new Liberal Government is currently undertaking cross country consultations to inform a National Food Policy and a Poverty Reduction strategy. The Liberals, in opposition when the Special Rapporteur was in town, were clearly listening.

Mission to Mexico

Of course for reasons of national sovereignty, no government likes to be told by outside sources what it should or should not be doing. However when an invitation has officially been issued there are norms of common courtesy and diplomacy in terms of how best to respond, especially if the advice offered is not necessarily to your liking. These norms are well demonstrated by Mexico, the only other OECD member state to been visited by Olivier De Schutter. He made two visits, the first in 2011 with a follow up review two years later. Broadly speaking even though the issues confronting food policy and food security could be regarded as more pressing than those in Canada, the visit followed a similar procedure and scope but with a much different and forward looking outcome.

The Mexican Government reported that the Special Rapporteur's second visit 'had been instrumental in facilitating the completion in October 2011 of the constitutional reform process which enshrined the right to food in the national constitution' (OHCHR, 2013b) and indicating that preparation of a draft framework law on the right to food was underway. In other words a highly constructive and positive response even though the questions raised were as challenging as those posed in the visit to Canada. In 2017 the right to food has been embodied in the constitution of Mexico City.

Mission to Poland

The other high income country, and OECD member state, to receive a visit has been Poland. The mission was conducted in 2016 by the current Special

Rapporteur, Hilal Elver (HRC, 2016). It followed a familiar right to food format exploring issues from the perspectives of food availability, food access (refuges and inmates), food adequacy and vulnerable and marginalized populations (rural women and women farmers, the Roma population and seasonal migrant workers). Its recommendations included the need to ratify the *Optional Protocols* of the *ICESCR* and *CRC*; devising and adopting a national law on the right to adequate food informed by gender-based perspectives and collecting disaggregated data to monitor the situation of all marginalized groups including women; and addresses issues of social assistance.

Interestingly, however, when commenting on food waste as an emerging issue the report noted the important role played by food banks and their links with food retail chains including Tesco Poland, Auchan and Carrefour. It expressed concern that 'it is not possible to donate and distribute food that has passed its "best before" date' regulation thereby preventing increased donations from partners such hypermarket chains to food banks (HRC, 2016).

What is disappointing is that the report refers to food banks as non-governmental organizations, not charities. Moreover, despite the fact that the Special Rapporteur argued correctly in her presentation to the Committee on Food Security in January 2017 that the right to food is not about charity, in Poland she has endorsed the alliance between Big Food, charitable food banking and surplus food as a solution to food waste and food poverty. Equating corporately dependent food charity with the right to food is problematic and urgently requires clarification.

Valuing rich world country missions

The importance of country missions to high income countries under the umbrella of the OHCHR is an emphatic statement that all UN State Parties to the *ICESCR*, rich and poor alike, have shared obligations under international law progressively to realize the human right to adequate food. Only the USA is an outlier. However, it is important to note that prior to Hillary Clinton stepping down from the Secretary of State portfolio in 2011, at the request of the Special Rapporteur, talks had been under way for him to conduct a country mission. Indeed, immediately before her departure there had been a positive response from the State Department but by then De Schutter's visit to Canada had been confirmed and it was felt unwise to conduct two missions to North America at that time, and priority was given to Madagascar and Malawi (ODeS – email communication 27.6.2017).

Whilst the context of SRRF country missions is that of advancing national food security the key driver is promoting and mainstreaming the human right to an adequate standard of living for all, inclusive of the most vulnerable. Compliance and taking action to address violations of the right to food apply equally whatever the wealth, GDP and status of the nation might be. Symbolically country missions hold out the promise of critical and collective solidarity.

The right to food is not intended as a one way message to the poor world. It also sends an inconvenient reminder to the affluent 34 OECD food bank nations which have ratified the UN *ICESCR* that democratic accountability, good governance, human rights, the rule of law and the role of public policy are of primary significance when millions of their peoples are dependent on corporate charity, enduring the stigma of eating surplus and wasted food. It is an expression of collective solidarity which recognizes that the market coupled with a dysfunctional food system and broken social safety net is no guarantor of human dignity nor of food security for all.

Given the relatively short life of the UN Special Rapporteur's office and the small number of OECD countries visited it is perhaps too early to judge the success of this mandate to advise on and monitor the national implementation of the right to food in the high income world. In large measure it depends on political will and the openness to external review of constitutional arrangements and the public accountability of the state.

However from the perspective of collective solidarity and human rights the fact that such country missions take place in the rich world ensures that the right to food attracts the political attention of governments. Yes, there are challenges as shown by the rude rebuttal of the Special Rapporteur by the former Canadian Government. Yet there are opportunities to advance the conversation in light of the process leading to entrenchment of the right to food in the Mexican Constitution. Not that it should be at the expense of the Global South but there is a strong case to increase the number of country missions by the Special Rapporteur to high income states, especially those which have become food bank nations with expanding food safety nets and secondary food markets. If we are willing to preach the right to food to the poor world perhaps we should first get it right in our own.

Monitoring the right to food: Periodic Reviews

Within the United Nations there are two processes whereby the right to food may be formally reviewed: the Universal Periodic Review conducted under the auspices of the Human Rights Council and the periodic State Party Reports (SPR) of the *ICESCR* undertaken by the Committee on Economic Social and Cultural Rights.

HRC – Universal Periodic Review

The Universal Periodic Review (UPR), adopted by the UN General Assembly in 2006, and conducted every five years, is 'a unique process' involving 'the review of the human rights records of all UN member states' (OHCHR, 2017c). It is a comprehensive review in which UN member states report on the actions they have taken to improve their record and fulfil their obligations regarding all aspects of civil and political as well as economic, social and political rights. It is a massive undertaking but an international mechanism which includes or should include the monitoring of the right to food.

Missed opportunities

However the opportunity the UPR offers to bring the right to food to the attention both of the Human Rights Council and the governments of OECD countries is at best limited if not bleak. For example, and granted the sample is small, there is no reference to domestic hunger or food poverty let alone the right to food in the national reports for Belgium, Canada, Finland, Germany and the UK (OHCHR, 2017d). There is some discussion of poverty reduction, welfare reforms and social security but no mention of charitable food banking and expanding food safety nets.

Perhaps this not unexpected given the totality of the human rights under review. With the exception of Germany and Belgium the civil society human rights community also appeared to be silent about the right to food. For the same reason this can be well understood in light of the issue of resource constraints and lack of person power.

FIAN Germany

It is therefore notable that FIAN Germany (2013) in its submission to the UPR process addressed the way food banks reinforce social exclusion as well as the issue of the privatization of the right to food while the state failed to reduce dependency on food banks. Taking into account the dramatic rise of charitable food banking it also made specific recommendations regarding Germany's obligations to implement the ICESCR regarding the right to food (FIAN-D, 2013; HRCWG/UPR 2013).

Disappointingly, however, as Ute Hausmann, the former director of FIAN Germany has commented 'in the end it was not taken up by the Human Rights Council so there was no reply by the government' (UH – personal communication, 24.6.17); and sadly, as Phillipp Mimkes, the current FIAN Germany director advised, there is no plan to bring the food bank issue forward in the current UPR process (PM – personal communication, 25.6.17)

FIAN Belgium

FIAN Belgium as a member of the Coalition for the Right to Food (CDA) also made a joint submission to the 2015 UPR of Belgium. It recommended that Belgium in order to meet its obligations should entrench the right to adequate food in framework law (CDA, 2015). It argued that 'the lack of a legal basis for the right to food makes it hard to turn to the law to ensure access to sufficient food for vulnerable peoples'; and recommends that the Government should make food aid a legal obligation (HRC-B, 2016a). However the Belgian Government turned down these proposals noting that the right to food was addressed by various laws and social measures in the country, and it was not interested in a generic framework law (HRC-B, 2016b). The submission also heard that a bill recognizing the right to food was in process and that Belgium should adopt it. Yet this was a bill proposed by the Greens who were in opposition and lacked any possibility of political support (ODeS – email communication, 27.6.17)

USA

Given the overall indifference by the OECD's wealthy UN member states to the right to food in the UPR reports, it is interesting if not ironic that access to food (along with health care) *is* specifically mentioned by the USA national report (HRC-USA, 2015a). It cites its 2014 investment 'of more than $103 billion in domestic food assistance programs, serving one in four Americans during the year'. The beneficiaries included 46.5 million individuals per month receiving food stamps (SNAP); 8.3 million individuals per month in receipt of the Special Supplemental Nutrition Program for Women and Children; 30.3 million children accessing school meal programmes; and 2.5 million elderly adults signed up to the Older Americans nutrition programmes. Additionally, from a food charity perspective, the report references the 785 million pounds of food received by emergency food providers through the Emergency Food Assistance Program; and concludes by stating there is 'substantial evidence that these programs improve social, economic and nutrition conditions for low-income Americans' (HRC-USA, 2015a, p.19).

This may be the case given the official US data on food insecurity (see Chapter 2). However there is no mention in the report of domestic hunger or food insecurity, the role of charitable food banking, nor of the right to food. The paragraph reads as a bold but in fact unsubstantiated claim highlighting the success of the publicly funded food safety net in reducing poverty. However, given that the USA is not a state party to the *ICESCR* and therefore does not participate in the CESCR monitoring process, it is welcome that the US is open to a UPR discussion about food access.

What is of more interest are the recommendations made by the expert HRC Working Group respecting the *ICESCR* comments and in particular the nature of the US Government's responses to them (HRC-USA, 2015b). A number were best summed up by Germany's comment that the US should 'take genuine steps towards the ratification of treaties and optional protocols to conventions which the United States has already signed, but not yet ratified' (# 4, p.12); and Cuba's that the United States 'guarantee the right by all residents in the country to adequate housing, food, health and education with the aim of eradicating poverty, which affects 48 million people in the country' (#309, p.32).

Of course these were the days of the Obama Administration which indicated its support for those recommendations requesting the USA to ratify treaties and conventions (p.8) and that while not being party to the *ICESCR* it understood that the rights therein are to be realized progressively (p.4). Some hope for the right to food? Perhaps if the Democrats had won the 2016 Presidential Election, there may have been further progress. It is however impossible to imagine the protectionist Trump Administration giving a moment's thought to ratifying the *ICESCR*.

Reflections

Despite the limits and challenges of the UN's mainstreaming of human rights and the political difficulties of monitoring the rich world's compliance with the right to food through country missions and the UPR process, it is vital that these international mechanisms are universally recognized and have been established in international law.

However indifferent or even opposed powerful states as 'primary duty bearers' may be to the global human rights agenda these international institutions serve as democratic and collaborative sites for advancing debate, fostering education and raising awareness about the importance of economic, social and political rights in general and the right to food and public policy in particular. In the struggle against domestic hunger in the OECD food bank nations they are practical expressions of critical and collective solidarity creating opportunities for civil society to remind us we are all 'rights holders'; and of the urgent need to change the conversation from corporate food charity to the right to food and in the process holding the State publicly to account and reclaiming public policy.

10

CIVIL SOCIETY WITH A RIGHT TO FOOD BITE

Reclaiming public policy

Prior to speaking at a Food Thinkers Forum at City University's Centre for Food Policy in London in 2014, I was asked by Tim Lang in a three-minute interview about the role civil society could be playing in addressing the food poverty and food charity issue. I suggested engaging 'the food policy movement with the anti-poverty movement with public health activists and actually politicize the question of charitable food banking because unless we understand this issue as one of income which is absolutely necessary we are going to see the permanent entrenchment (of food banking) in this country (the UK) over the years to come; and that is going to be incredibly problematic in dealing with this problem of food inequalities' (FB, 2014).

With an extra minute I would have added that domestic hunger in a country as wealthy as the UK, indeed in all OECD food bank nations, is fundamentally a matter of economic, social and cultural rights long neglected by the neo-liberal State. Civil society including academia are crucial agents not only for putting politics back into hunger but also for mobilizing human rights, particularly the right to food.

Such an assertion requires justification in that the FAO *Right to Food Guidelines* not only recognizes the State as the '*primary duty bearer*' for realizing the right to adequate food but promotes a multi-stakeholder approach to national food security 'encompassing civil society and the private sector' (FAO-VGs, 2005). The problem is the private sector and the indifferent State hold the cards. While corporate food charity continues feeding the need, governments with the power to address the structural causes of domestic hunger and food insecurity choose not to act.

From the standpoint of critical and collective solidarity and reclaiming public policy I am therefore thinking of civil society, as Cresswell Riol has suggested, with a more inclusive meaning which recognizes and values all those with whom it works and represents as rights holders and participants in the struggle (see Cresswell Riol, 2017).

In that sense in the discussion which follows civil society refers to leading international and national food justice CSO's such as FIAN International, Nourish Scotland, Food Secure Canada, UK Food Research Collaboration and the embryonic Right to Food Coalition in Australia and their partner organizations. Their advocacy as participants in monitoring and holding their governments to account for complying with their obligations under international law to realize the right to food is painstaking and immensely challenging given widespread and influential resistance to 'rights talk' in the rich neoliberal world. It is work which has to be undertaken with a bite.

'Rights' talk

Of course many believe there is no need for the right to food let alone using the language of human rights and international law. 'Rights' talk is just that: all talk and no action, soft law, voluntary guidelines, solidarity in disguise, a nod in the direction of the poor and little more. Hunger is best left to corporate philanthropy and the food bank industry. After all, in its own language food banks are the link between waste and hunger. In any event real hunger does not exist in the rich world when compared to malnutrition and starvation in the desperately poor countries and fragile or failed states of the Global South.

Yet as Louise Arbour has argued 'the reason that "rights talk" is resisted by the powerful is precisely because it threatens (or promises) to rectify distributions of political, economic or social power that, under internationally agreed standards and values, are unjust' (Arbour, 2005). This is precisely the reason why human rights matter and why it is crucial that civil society advocacy is indispensable in monitoring the national compliance of the rich world's UN and OECD member states with the *ICESCR* and the right to food.

In a neoliberal driven world for all practical purposes the right to food is missing from political debate in wealthy OECD food bank nations. The normalization of Foodbanks Inc has obviated any such need – hunger is being solved. In any case 'rights' talk threateningly conjures up visions of the Nanny State, a culture of entitlement for the poor, higher taxes and the progressive distribution of income and wealth. Much better to leave domestic hunger to food charity and corporate social responsibility. The right to food is the path not taken.

In the USA, the right to food certainly has little traction. As Andy Fisher has noted while it is an important educational and organizing tool, it has two key limitations: 'the challenges of communicating economic rights in a political environment that seeks to reduce or eliminate existing government entitlements' (an argument that would likely hold in the majority of OECD members states), and 'its focus on the government as the sole guarantor of rights [which] misses key opportunities for ensuring the public's food security' (Fisher, 2017, p.38). He rightly points to citizen pressure directed not at the state, but at corporate power such as boycotts and consumer education, which has improved the lives of millions.

Yet as Alison Cohen, Senior Director of Programs at WhyHunger in New York has noted 'the framing of the right to adequate food and nutrition could serve as a tool for developing a new counter-narrative that shatters the perception that hunger in the U.S. can be solved by charity, increasing food production, and market forces' (AC – email communication, 10.10.2017). The point is human rights based approaches (HBRA) are central to the struggle for poverty reduction achieved through collective solidarity and reclaiming public policy. It is the legislative power of the State which can make this happen.

The right to food is not just talk. It is a legal and justiciable concept. Moreover, its implementation benefits from a framework of international law, instruments and mechanisms rooted in universality and critical solidarity with the poor and hungry. They inform and enable the development of comprehensive strategies for moving beyond dependence on Big Food, Big Philanthropy and charitable food banking to legislative action, policies and programmes and judicial remedies respecting human dignity. The challenge is how best to make use of international human rights law and its institutional mechanisms and resources it has at its disposal.

Rich world compliance with the right to food, or not

The international mandate and specific mechanism for monitoring member states' compliance with the right to food in the context of the *ICESCR* rests with the UN Committee on Economic, Social and Cultural Rights (CESCR, 2017) based in Geneva. Established in the early 1990s it predates the more comprehensive human rights UPR process by more than a decade with 164 state parties, excluding the USA, having ratified the *ICESCR*.

CESCR monitoring: State Party Reports (SPR)

Governments are required to submit periodic State Party reports every five years indicating their compliance with ESCR obligations including the right to food as a key component of the right to an adequate standard of living (Article 11) (see CESCR, 2017). Meeting in Geneva, the Committee, comprising 18 independent experts, considers each State Party's report and then engages an iterative process for assessing the progress made, or not, which includes the participation of civil society presenting parallel or 'shadow' reports.

Following the CESCR's initial examination a List of Issues (LOI) is submitted to the respective State Party requesting further information and or comment. The government's reply to the concerns expressed (RLOI) are then further considered by the Committee which provides Concluding Observations (CO) to the government noting progress made (compliance) and recommendations in the case of violations and where attention is needed (see Adzakpa, 2016). The COs are referenced and used as benchmarks in future periodic reviews thereby reminding State

Parties of previous concerns, thus providing a historical record of each government's progressive compliance, or not, with its ICESCR obligations. This monitoring process is intended to be consultative, transparent and accountable.

Procedural issues: rhetoric and reality

However as Hannah Adzakpa has pointed out in a meticulous study undertaken for Nourish Scotland, the national food policy and food justice CSO, all is not as it seems. EU countries reporting on their implementation of the right to food in the CESCR Periodic Review process leaves much to be desired. As she comments while states claim 'to fully implement their obligations, the reality is very different' (Adzakpa, 2016, p.4).

The study, comparing the most recently submitted EU state party reports (SPR) between 2000–2016, first highlights a number of procedural weaknesses including governments generally being late in submitting their reports, and not being relied upon to follow the CESCR reporting guidelines (pp.20–21; p.4). Of the OECD member states within the EU, the UK won the prize for being the only state party to meet its reporting deadline in 2014. The lateness of other EU (OECD) states stretched from one month (Sweden) and 5 years (Ireland), 8 years (Belgium) and even 12 years (Latvia), and then took two to three years before being reviewed (Adzakpa, 2016, table 4). This is not simply an EU phenomenon but one shared by other OECD member states. For example the most recent SPRs of Australia, Norway and Japan were each two years late; Canada's three years and Mexico's four years behind schedule.

There is no doubt that on administrative grounds alone the CESCR periodic review mechanism for holding countries to account faces considerable challenges when it comes to reporting on the right to food when taking into account the full slate of economic, social and cultural rights under review. The procedures are lengthy, data collection complex, many reports are long overdue and yet, the iterative process is thorough and one which invites civil society participation. It can be argued that the Concluding Observations do hold governments publicly to account for meeting their ESCR obligations under international law.

In that sense a real strength of the CESCR SPR process is the regular opportunity it provides for monitoring the extent to which UN member states including the OECD food bank nations have embraced the right to food and its explicit obligation to ensure the adequacy of employment incomes and social security benefits enabling all to purchase the food of their choice. Broadly speaking it casts an international light on national framework legislation on food policy and the development and the implementation of 'joined-up' food policy national action plans directed at the optimal nourishment of the population and the eradication of poverty.

In this process the UN mechanisms serve as a constant reminder that government is the 'primary duty bearer' for ensuring that the right to food is '*respected*', '*protected*' and '*fulfilled*', and that domestic hunger as a profoundly political issue must not be left to charity. It means connecting the policy dots to achieve shared economic prosperity, social justice and a sustainable food system underpinned by international human rights law. While the State (and the judiciary) holds most of the cards it falls to civil society to make sure they play them. The task is putting the right to food to work so as to reclaim public policy. It is about putting the politics back into hunger thereby giving voice to the needs of vulnerable populations. It is a process facing considerable challenges.

The non-compliant indifferent State

This is why it is so crucial to monitor the compliance of the wealthy food bank nation states with their obligations. It is not simply an administrative issue of being years behind in submitting their reports (though meeting the required 'best before date' would strengthen confidence in the CESCR review process), it is much more a question of substance and transparency in keeping faith with the reporting guidelines. For example, even though each member state is required to report separately on the right to food, they may or may not; they pick and choose which of the normative core content – availability, adequacy and accessibility – they will comment upon; and of particular concern 'they rarely comment on problem areas such as the affordability and financial accessibility of food' (Adzakpa, 2016, pp.20–21; p.4). In terms of domestic hunger and poverty this last concern is at the heart of the right to food debate, indicating non-compliance and a long way to go.

EU/OECD food bank nations

It was with commendable efficiency that the UK submitted its 6th Periodic Review on time, but nevertheless the report failed to comment on the matter of financial accessibility and food's affordability for those standing in the breadlines. Six EU states – France, Germany, Greece, Italy, Latvia and the Netherlands made no mention of the right to food though eleven did report separately on it: Austria, Belgium, Denmark, Finland, Hungary, Ireland, Latvia, Portugal, Slovakia, Slovenia and Poland; and others within the broader context of *ICESCR* Article 11 (ibid, p.21). Even when State Parties did comment on the right to food, evidence-based data was not necessarily offered supporting their claims that the right food was being met. Rather food's availability, its safety and nutritional adequacy and awareness raising would be vouched for in the reports creating the impression that financial access to food was assured for vulnerable populations.

Significantly, given widespread food insecurity and growing dependence on charitable food banking and secondary food safety nets, Adzakpa noted that the CESCR's Concluding Observations directed the attention of Austria, Ireland and the UK to the matter of 'the financial *inaccessibility* of food' (2016, p.28). In plain English, charity was an insufficient response to food poverty and food access. Income-based policies were required if domestic hunger was to be successfully overcome. As research tells us food insecurity is a symptom of poverty, inequality and broken social safety nets. Unless and until these structural determinants are acknowledged and acted upon no wealthy food bank nation can legitimately claim or imply that it is progressively realizing the right to food and assuring food security for its vulnerable populations.

Non-EU/OECD food bank nations

There was a similar lack or pattern of ambiguous reporting by the non-EU food bank nations. While all OECD countries commented on social security (Article 9) and the right to an adequate standard of living (Article 11), the right to food was a hit and miss affair. Australia, Chile, Korea, Iceland and New Zealand included no such reference, nor did Switzerland though it mentioned the *Voluntary Guidelines*.

Canada's 6th SPR for the period 2004–2009 showed no understanding of the right to food. The Federal Government and half the provinces and territories including Ontario, the largest province, made no mention. The remaining provinces and territories made passing reference but with Quebec and other provinces presenting food banks as an acceptable response to food insecurity. While attention was given to social security and the right to an adequate standard of living it lacked reference to official food insecurity data and financial access to food. Indeed by the time of the final CESCR discussions on Canada's Report in 2016, it was already seven years beyond its 'best buy date' (CAN, 2013).

Israel, Japan, Mexico, Norway and Turkey referred to the right to food but none fully addressed the relationship between income inadequacy and food insecurity (OHCHR, 2017e). Even the Norwegian SPR which recognized the *Voluntary Guidelines* failed directly to discuss the matter of food access in terms of financial inability (NOR, 2012). The general assumption seemed best expressed by Japan's SPR which noted that 'an appropriate food supply has been achieved in Japan and the rights of certain deprived groups to receive an adequate supply of food are not being infringed' (JAPAN, 2011, p.51). In other words, food banks were not regarded as concrete evidence of financial inability to acquire food; and a simple assertion that the food supply is adequate, relatively safe and nutritious (despite high rates of obesity) somehow means access to food is not a problem. How can this be when all these State Parties are members of the rich world food bank nations club?

Hannah Adzakpa's comment is worth repeating: 'EU member states' (I would add the majority of OECD states) 'must stop treating the right to food as a purely aspirational idea or an obligation that only applies to developing countries. Rather, they must take it for what it is: a human right, the normative content of which must be protected and progressed – even in times of resource constraints' (2016, p.4). No matter whether an economy is growing or experiencing austerity, collective solidarity expressed through the right to food must be the first priority.

USA: absentee food bank nation

Ironically the USA, as the birthplace of modern day charitable food banking, is the one OECD country unbothered by having its right to food record subject to monitoring by the CESCR. Certainly, the USA participates in the more general human rights UPR process, but by pleading sovereignty and not ratifying the *ICESCR* the USA remains immune to the focused international monitoring of its welfare state and its compliance with the realization of economic, social and cultural rights. Widespread domestic hunger, food insecurity, income inequality, poverty and the role played by public food assistance programmes and corporate food charity avoid international monitoring.

Nor is this likely to change any time soon especially when President Trump's appointee, Ben Carson, the new head of the Federal Department of Housing and Urban Development has just declared that 'poverty is a state of mind' (NYT, 2017). Whilst the USA's food (in)security data is to be respected it is also a country which lacks universal health care.

Yet it should be noted certain American food justice organizations such as WhyHunger, Bread for the World and Witnesses to Hunger advocate on behalf of the right to food; and as Molly Anderson makes clear a wide range of community-based and food justice programmes exist 'for low income people that help them grow their own food, get more affordable access to fresh fruits and vegetables and teach them how to cook for themselves and make healthier food choices'. Yet what is missing in the USA, she argues is 'focused attention on the root causes of hunger and food insecurity' (Anderson, 2013).

She argues what is really needed is 'a true living wage, employment for all who are able to work, reducing corporate influence on the US government and pernicious manipulation of food choices, and creating a mandatory social safety net for everyone should be considered as means of combating hunger in the United States'. However, Anderson contends the most serious missing piece is 'concerted effort to hold the US government to account for the realization of the right to adequate food for all people', noting that 'community-based initiatives will continue to serve some portion of the US population whose right to food is violated; but it will never substitute for recognition of the right to food' (Anderson, 2013). In that context it is important to recognize that WhyHunger has been working with the FIAN-based Global Network for the Right to Food and Nutrition (GNRtF) looking at the 'false solutions to hunger that are spread by multi-national corporations' (Alison Cohen – email communication, 23.10.2017).

These are powerful messages not only within the USA but in all OECD food bank nations which having ratified the right to food continue to ignore it. The tardiness and incompleteness of their five-year periodic SPRs respecting the right to food hardly demonstrate political will in addressing domestic hunger. Indeed how concerned are they? Their welfare states and publicly funded systems of income security and social protection, however well established, are steadily being eroded by the global expansion of US style charitable food banking straight from the heartland of neoliberalism, the most powerful nation state itself not subject to formal international monitoring and account.

Corporate food charity has proved ineffective in the USA for addressing food poverty, all the while undermining the right to food and income-based approaches to social protection and poverty reduction. Yet this business model of charitable food banking has since the early 1980s become entrenched in the OECD world, and beyond.

Civil society advocacy: holding the State to account

Addressing domestic hunger from the perspective of food access and financial affordability, it is highly regrettable that the US Government declines to sit at the UN's right to food table and, likewise, that many OECD governments emboldened by neo-liberalism and the false promises of corporate food charity are indifferent and reluctant participants in the CESCR Periodic Review process. However, it is extremely encouraging that civil society has a formal and indispensable role to play as participants, consultants and advocates in holding national governments to account.

Parallel and 'shadow' reporting

JustFair, the England based CSO which acts as a monitoring and advocacy agency for the protection of economic and social rights, has rightly commented 'the (periodic) review is an invaluable opportunity for civil society organizations to make the case for justice and human rights in the UK' (JF, 2016, p.1). This includes submitting parallel or 'shadow' reports, providing 'information in response to the Government's own data' with the opportunity to influence the Concluding Observations made by the CESCR. JustFair emphasizes the 'vital role' played by civil society in ensuring that the national data being reviewed is accurate and that the state is complying with its international human rights obligations including the right to food.

Additionally as JustFair also notes the resultant Concluding Observations provide first rate advocacy material for CSOs. It stresses the point that involvement in 'the review process is therefore an excellent way of **holding the government to account and making a positive impact** on people's lives' (JF, 2016, p.2, *bold in original text*). The formal participation of civil society in the UN's monitoring of ESCR rights is further evidence that the right to food matters, underlining the critical role CSOs play in changing the conversation about domestic hunger from charity to rights. It validates the belief that we are all rights holders; that the right to food has legs and putting it to work is directed at reclaiming public policy.

Advocacy at work

Important examples of how civil society engages the five-year CESCR Periodic Review process in order to incorporate the right to food into public policy are found in many food bank nations. At the United Nations, FIAN International plays a crucial role in this process working in many countries including within the rich world. In the EU the case of Germany merits attention (Hausmann, 2013).

Examples from other OECD member states include the work of leading food policy CSOs in Canada, Scotland and Mexico which suggest that over time governments may well be listening. In Australia right to food action is in its infancy and as noted earlier even within the USA rights-based approaches to domestic hunger are to be found.

It must be emphasized that in the CESCR Periodic Review process, food policy NGOs frequently needs joining up work in collaboration with legal rights, anti-poverty, social justice and environmental organizations to address food poverty and wider issues of economic, social and cultural rights. At the same time the task of successfully influencing government does not depend on presenting parallel or shadow reports every five years. Raising awareness and public education also benefit from country missions and conference participation by the UN Special Rapporteur and by the FAO RTF Team. However it is the active HRBA to research, campaigning and collaborative action by local and national food, public health, environmental, social justice, anti-poverty and human rights organizations over time which are the keys to successful advocacy. The CESCR Periodic Review process inclusive of civil society provides a clear focus and regular checkpoints for assessing State compliance with the progressive realization of the right to food.

FIAN International

Historically, the success of bringing transparency and democratic accountability to the CESCR Periodic Review process itself could be attributed to the tireless work of FIAN International, the global NGO human rights organization which has worked on realizing the human right to adequate food and nutrition (see FIAN, 2017). With roots in Amnesty International (AI) and its focus on civil and political rights, FIAN was founded in 1986 as 'FoodFirst Information and Action Network – FIAN' to advance economic and social rights with an important focus on securing 'people's access to the resources that they need in order to feed themselves, now and in the future' (FIAN, 2017a).

Anita Klum, FIAN's president has written that back in the 1960s and 1970s (AI was founded in 1961) 'ESC rights were not considered justiciable: they were not regarded as rights but as a sort of "being kind to poor people"' (Klum, 2016). Looking back over 30 years since its founding FIAN, with national sections in more than 50 countries worldwide, including Belgium and Germany as original OECD states, has adopted AI's model of work 'urgent actions, case work, as well as advocacy activities and campaigns' (Klum, 2016).

Significantly in 1989 FIAN was granted consultative status to the UN. This placed it in a prime position to support and advocate with and on behalf of its national sections in bringing forward 'parallel reports' to the CESCR Periodic Review process, including Spain in 2004 and Austria, Belgium, Germany and Norway in 2013 (FIAN, 2017b). These 'parallel FIAN reports from OECD countries', as Anne Bellows, professor of food studies, has noted, 'are drafted largely by the FIAN country offices as opposed to FIAN International', also advising that FIAN works collaboratively within the Global Network on the Right to Food (GNRtFN), playing the 'arbiter' role and similarly with its publication, the *Right to Food & Nutrition Watch*. Anne Bellows also notes that FIAN plays a leading role internationally regarding the corporate cooptation of government decision-making in food systems and public mandates generally. These would include the corporate capture of emergency food systems and food banks (AB – email communication 8.6.17).

Food Secure Canada

Of the 33 OECD countries which have ratified the right to food (*ICESCR*), Canada holds the longest track record as a member of the rich world's food bank nation club, in fact a co-founder with the non-ratifying USA. Ironically, following closely on the heels of the Nordic countries, it was one of the earliest OECD states to ratify the *ICESCR* in 1976, followed a day later by the UK. Those it can be argued were the heady days of the welfare state before the steady drip of neo-liberalism and growing corporate power began to undermine public policy informed by economic and social rights.

In 1993 the CESCR's Concluding Observations in response to Canada's 2nd Periodic Review expressed concern about the persistence of poverty in Canada, observing: 'there seems to have been no measurable progress in alleviating poverty over the last decade, nor in alleviating the severity of poverty among a number of particularly vulnerable groups'. It even went further stating its concern about the '*evidence of hunger in Canada and the reliance on food banks* (author's italics) operated by charitable organizations' (E/C.12, 1993).

A quarter of a century later Canada's domestic hunger still remains a human rights concern. In 2016, the COs to the 6th Periodic Review while welcoming the possibility of a national food policy, reiterated its long held concerns about 'the rates of food insecurity in the State party, *the increased reliance on food banks* (author's italics), particularly in Northern Canada, and the deficiencies of the Nutrition North Canada food programme' (E/C.12, 2016). From the perspective of civil society holding government to account, what then has changed?

As for access to food, very little it appears. However it could be said that today civil society may be seeing the fruits of its advocacy as it witnesses important policy developments by the federal government. Food Secure Canada, the leading national food justice and policy organization, has been playing the long game. In the late 1970s the People's Food Commission, a citizen led initiative which held 75 hearings across the country released its report *The Land of Milk and Money* arguing for a national food policy which also drew attention to questions of food accessibility (PFC, 1980).

In the mid 2000s the idea of revisiting the need for a national food policy resulted in a second report '*Resetting the Table: A People's Food Policy for Canada*' (PFPC, 2011), produced in collaboration with Food Secure Canada. Recognizing that the federal political parties and key sectors in the food industry were advancing ideas for new national food policies, the People's Food Policy was grounded in the idea of food sovereignty and the need 'to build a healthy, ecological and just food system for Canada' (PFPC, 2011, p.1). Amongst its many recommendations it argued for 'guaranteed universal access to healthy and safe food' and 'the implementation of a federal poverty prevention and elimination strategy' (p.22).

Still, not unexpectedly, the Stephen Harper led Federal Government refused to listen. It took the 2012 visit of UN Special Rapporteur Oliver De Schutter to shake the branches. He reported that national food policy and framework legislation were not to be found and, moreover, that food banking was an inadequate response to domestic hunger. In light of his abrupt and un-Canadian dismissal by Ministers of the Crown, the Special Rapporteur's message had no immediate effect. However the progressive work of Food Secure Canada under the leadership of its newly appointed director, Diana Bronson, continued apace as did the measurement of the prevalence of food insecurity undertaken by Valerie Tarasuk and her team at the University of Toronto's PROOF research centre.

In 2015 the Federal Government changed and Justin Trudeau's Liberal Party took power. By no means did this herald a mandate for rights-based approaches to national food policy nor social policy incorporating income-based solutions to domestic hunger. However it did encourage a new flow of ideas and debate about evidence-based and forward looking public policy. Fortuitously, the change in Ottawa coincided with Canada's 6th State Party Report (CAN, 2013). Together with other leading national NGOs including the parallel reports of Canada Without Poverty and the Social Rights Advocacy Centre, Food Secure Canada ensured that in 2016 the CESCR's Concluding Observations brought food insecurity and the right to food to the attention of the Federal Government, and particularly the crisis of food insecurity in Northern Canada. The need for national framework legislation and the justiciability of economic, social and cultural rights were addressed as pressing concerns.

It could be that the political tide has turned in favour of reclaiming public policy. In early 2017 the Government of Canada launched an on-line and in person consultation across the country to develop a *national poverty reduction strategy*. It is an inclusive process including setting up an advisory committee to the Minister of Families, Children and Social Development bringing together those who have experienced poverty as well as a wide range of experts with basic income programmes up for discussion. A few months later the Minister of Agriculture and Agri-Food Canada introduced an online survey aimed at developing a *national food policy*. It is focused on: increasing access to affordable food, improving health and safety, conserving soil, water and air and growing more high quality food. it would seem the voice of civil society and the international community has been heard.

While there is as yet no recognition of the right to food in either of these federal consultations, it will continue to be brought to their attention by Food Secure Canada's continued advocacy for a people focused national food policy based on *'Five Big Ideas for a Better Food System: The Right to Food; Healthy and Sustainable*

Diets; Sustainable Food Systems; Food and Reconciliation; and *More Voices at the Table'* (FSC, 2017). As FSC argues 'we currently have a patchwork of agricultural policy, health policy, environmental policy, social policy and so on. In order to create a better food system, we need to start thinking more comprehensively about how we govern our food, from farm to fork'. From a food access perspective, connecting the policy dots between the Federal Government's proposed national poverty reduction and national food policy strategies remains to be seen.

Nourish Scotland

Across the Atlantic Nourish Scotland (NS), the leading NGO campaigning on food justice issues in Scotland, works collaboratively with other organizations 'from trade unions to conservation NGOs', and offers an instructive example of effective advocacy to advance rights-based approaches to poverty reduction and access to food for low income populations. NS's advocacy for the right to food, similar to that of Food Secure Canada, seeks to ensure: the availability and supply of food; its nutritional adequacy and safety; and financial and geographic access (Kontorravdis, 2016).

As Pete Ritchie, Nourish's Executive Director has explained the right to food dialogue gathered steam as the result of two gatherings held in 2014 and 2015, first the NS international 'Commonwealth of Food' conference, followed by the 'Beyond the Food Bank' symposium organized by the Church of Scotland. NS realizes that successful action requires broad-based support across the NGO community. This is being achieved through the Scottish Food Coalition (with its 18 NGO partners), co-convened by NS and the Royal Society for the Protection of Birds, an indication that access to food is not only a pressing issue for *homo sapiens* but for all species! (PR, personal communication, 6.6.17).

Elli Kontorravdis, a lawyer by training and NS's Policy and Campaign Manager, has outlined the further steps taken along the right to food way. Firstly she points out that the Scottish Food Coalition of which Nourish is a founding member 'is pushing for rights-based food governance structures which incorporate and progress the right to food through new legislation'. In March, 2017 the Special Rapporteur on the Right to Food, Hilal Elver, was invited to visit. She took the opportunity to 'catalyse the conversation further for the Scottish Government by quite publicly saying that Scotland could be a European leader by adopting such an approach. Both European and leadership are values the Scottish Government identifies with closely' (*sic*). Significantly, a consultation on legislation is scheduled to begin in late 2017, aiming to introduce a 'Good Food Nation' bill with cross-party support (EK, email communication, 5.6.17).

Secondly, in terms of access to food, 'the Menu for Change Partnership – between Nourish, Oxfam Scotland, the Poverty Alliance and the Child Poverty Action Group', also with cross-party support, was launched in 2016. This grew out of the Scottish Government's Independent Working Group on Food Poverty which in 2016 published *Dignity – Ending Hunger Together in Scotland,* a report directly relevant to a rights-based approach to resolving food insecurity. The plan of action, already underway, is for a 3-year intensive action learning programme with three local authorities – East Ayshire, Fife and Dundee – making full use of their powers to address the underlying causes of food insecurity (EK – email communication, 5.6.17).

A third relevant group has been the Human Rights Consortium Scotland of which Nourish has become a member. It operates as a general rights-progression group coordinating civil society evidence and participation at various UN treaty-body committees, and collectively advocates for relevant follow up to the Scottish Government. Nourish's participation has enabled it to better embed the right to food into the priorities of the wider human rights network in Scotland (see EK, email communication,6.6.17).

It is clear Nourish's broadly based and collaborative approach to advancing the right to food dialogue is on the right track. However explaining what this means in practice not only with civil society but also government is demanding work. It also requires pursuing its advocacy agenda in the international arena. In this sense Nourish points to its success in holding the Scottish and UK governments to account. As part of the Scottish civil society delegation, Nourish met twice in Geneva in 2015 and 2016 with the CESCR Committee considering the UK Government's 6th State Party Report.

Nourish's parallel report, authored by Elli Kontorravdis, responding to the LOIs was presented during the formal process to the Committee of Experts. She noted that 'after the LOIs were released the Scottish delegation of NGOs continued to meet with the Scottish Government, which we certainly believe is one of the reasons why the Scottish Government's response to the LOIs is more purposive than that of the UK Government' (EK – email communication 7.6.17). As previously mentioned the CESCR Committee was very concerned about the failure of the UK Government (as well as Austria and Ireland) 'to ensure domestic implementation of the right to food' and its financial accessibility (Adzakpa, 2016).

The collaborative actions of Scottish civil society to building awareness about human rights-based approaches merits international attention. Roundtable discussions continue with the Scottish Government on how it can best meet its obligations as the '*primary duty bearer*' to 'respect', 'protect' and 'fulfil' the right to food – 'this is in stark contrast to the UK Government's response which has in our experience been complete silence' (EK – email communication,7.6.17). Thus far 'rights' talk has been effective but it remains to be seen how this is reflected in public policy, and south of the border in the UK parliament. What is significant is that the right to food is on the political table.

UK Food Research Collaboration

Interestingly as noted previously (Ch. 2) the UK's Food Research Collaboration which brings together academics and national civil society organizations working on interrelated food, environmental, health, social policy and human rights issues has stated publicly that the UK Government by not measuring household food insecurity is neglecting significant health consequences which 'are long term, severe and expensive' declaring 'it is time to count the hungry' (FRC, 2016).

The FRC strongly advocates that 'food insecurity should be routinely measured in the UK so we know who is affected and can target policy and resources on prevention, thereby avoiding unnecessary increases in health care costs' (FRC, 2016). Significantly the routine collection of such data is a necessary first step for informing evidence-based policy. Notably the FRC also published a Policy Brief in early 2017 critiquing the use of surplus food to feed hungry people while the UK Government continues to ignore the deep rooted problems of poverty (Caraher and Furey, 2017). While neither report explicitly adopts a right to food agenda their analyses are critical components of such an approach to public policy.

Australia's Right to Food Coalition

On the other side of the OECD world, Australia is in the early stages of building a Right to Food Coalition (RTFC) intent on improving 'the health and well being of all Australians by working to ensure equitable access to nutritious food'. It is a volunteer led coalition built from the grassroots from a broad spectrum of 'organizations, practitioners, researchers and community workers united in its cause' (RTFC, 2017). With its roots in a food systems group established in 2005, the RTFC was launched in 2016. It currently includes individual membership and over 50 organizations signed up. There are already local chapters in Sydney, NSW and the states of Tasmania and Victoria.

A principal concern is the Australian Government's non-compliance with 'its legal and moral obligation to guarantee the human right to adequate food for at least 1.2 million people without access to safe, affordable food and nutritious food' (RTFC, 2017). Its central message to government calls for affordable, accessible and healthy food being available to all, and a publicly funded social safety net for the most vulnerable (Mann, 2016). These demands mirror growing concerns being expressed in other austerity punished OECD member states whose governments for too long have ignored the widespread prevalence of domestic hunger and food insecurity. Significantly, however, the very establishment of the Coalition in Australia indicates that the UN's right to food discourse and the role of government as '*primary duty bearer*' is gaining community, professional and public attention.

In communicating with Liz Barbour and then speaking directly with Rebecca Lindberg, Liz Millen and Luke Craven (all Coalition organizers), it was quickly apparent that the RTFC means business but is hugely demanding of volunteer time and resources. In 2014, a conference 'Putting Food on the Table' was held in Sydney to set a future course of action. It was attended by over 200 delegates from different levels of government, civil society and academia working in areas such as food security and public health (obesity and income poverty), social work, social policy advocacy, legal rights and the environment. It included board and field staff from national and local level charities, food banks and rescue organizations. The success of the conference led to the call for the formal establishment of the Right to Food Coalition (RL,EM-M, LC skype communication, 5.6.17).

The need for the Right to Food Coalition is abundantly clear. As was explained the widespread nature of food insecurity is not seen as a problem in Australia, being described as a 'pretty barren' landscape with the right to food not yet on the official policy table. Evidence of this neglect is to be found in Australia's 5th Periodic Review (2014) which has no mention of it. Mind you other OECD countries show similar disregard. Discussion of food security is limited to a brief paragraph in the section concerning the impacts of climate change focused on the provision of licensed community stores to meet nutritional and related household needs in remote Indigenous communities in the Northern Territory (AUS, 2016, p.32).

Coalition members hold differing opinions about the role charitable food bank banking and its dependence on wasted and surplus food compared to income-based programmes of social security and the need to build the publicly funded social safety net. Again, these issues are not unique to Australia. It is nevertheless a complex and demanding task in a federal state to bring together a somewhat 'fractured food policy scene' which includes advocates for food sovereignty, corporate food banking and charitable food hubs in the context of a right to food approach. Added to this is the challenging task of advancing 'joined-up' food policy within national framework legislation informed by the right to food which requires the support of different levels of government: federal, state and municipal to say nothing of addressing the food sovereignty of Australia's Aboriginal peoples.

However the Coalition has plans to work on an invitation to the UN Special Rapporteur to conduct a country mission as a necessary step in further raising awareness about food insecurity and the right to food. Indeed, as Rebecca Lindberg and colleagues have written 'the human right to food, and the precedents for its successful implementation in dozens of countries around the world, is a powerful means by which to achieve the goal of a truly fair and just Australia in which everyone is well housed and everyone enjoys nourishing food every day. It's up to us to make this a reality. We can begin by reflecting human rights in how we deliver services, set policies and hold Australian governments to account' (Lindberg et al, 2016). Precisely!

Reflections

In light of widespread domestic hunger and the corporate cooptation of charitable food banking in the rich OECD world it is little wonder that the State Party Reports to the CESCR either omit or pay scant attention to the right to food. Unlike the comprehensive Universal Periodic Reviews where it could be argued that the right to food (however fundamental) is crowded out by the full range of human rights obligations and issues, there is no excuse when specifically considering the issue of food access and affordability in the context of economic, social and cultural rights.

If it were not for the advocacy of food justice CSOs including FIAN's national sections in OECD food bank nations such as Austria, Germany and Belgium, Just

Fair and Nourish Scotland in the UK, and Food Secure Canada together with their civil society partners working on poverty, housing rights, public health, social policy, civil liberties and environmental justice issues, governments would have free rein in continuing to look the other way. Nourish Scotland's research and civil society's submission of the parallel or 'shadow reports' play a vital role in exposing the studied indifference of national governments to domestic hunger as judged by the procedural lateness of many SPRs and their failure to connect the dots between domestic hunger, the right to food and public policy obligations.

It also matters that the USA, as the founding member of Foodbanks Inc which has spread across the rich world and beyond is not a State Party to the *ICESCR* periodic review. Not only does this prevent regular international scrutiny of the progressive realization of economic, social and cultural rights in the world's most powerful nation, it also weakens the capacity of the USA rights-based anti-hunger advocates and civil society partners to engage in the debate about the right to food with the Federal Government, thereby raising public awareness. By the same token in Australia it is important to recognize the beginnings of a right to food coalition to hold the Commonwealth Government to account.

It also needs stating that the periodic reviews conducted under the auspices of the UN Human Rights Council and the Committee on Economic, Social and Cultural Rights act as constant reminders to civil society in the rich world of the important role it can and must play in participating in this public policy process. Significantly it contributes to a transparent and historical accounting of State compliance or indifference to the fact that we are all 'rights holders'. Moreover, with the State as *'primary duty bearer'* it is crucial that civil society through its advocacy both monitors and makes demands on national governments regarding their obligations under international law in the struggle against domestic hunger and food poverty. In this way civil society plays powerful roles in reclaiming public policy and putting the right to food to work.

PART IV

Gathering political will

11

CHANGING THE CONVERSATION
Challenging propositions

The 35-year story of the global spread of US-style charitable food banking in the rich world's embattled welfare states is of promises to feed all those 'left behind' with surplus food, while its corporate partners benefit from neoliberal economic policies fuelling a prolonged crisis of income poverty and widening inequality. Food waste, domestic hunger and food banks are symbols of the moral vacuum at the heart of neoliberalism, symptoms of dysfunctional food systems and privatized and broken social safety nets.

It is particularly a story of indifferent rich world governments ignoring public policy and their obligations under international law to realize the right to food (*ICESCR*, 1976), and of abandoning the poor to the business of corporate food charity, in the process creating stigmatizing secondary welfare systems and secondary food markets for secondary peoples.

Foodbanks Inc and the indifferent State

It is also a story of deep irony. Since 1981 when food banks first crossed into Canada from south of the border, by 2015 all OECD member states had established food banks in one form or another. Many looked for inspiration to the USA as the home of food banking, and today benefit from membership in the European Federation of Food Banks, the US-based Global Foodbanking Network and Food Bank Leadership Institute. It must surely be noted that the world's richest and most powerful country with the highest levels of wealth and income inequality (DFI/ Oxfam, 2017) and with long established public and charitable food assistance programmes has never ratified the right to food. Meanwhile indifferent OECD countries with established systems of income security and publicly funded welfare states grounded in human rights seem not to care as they import and implant corporate food charity and food safety nets when there is little evidence of their effectiveness in combatting domestic hunger.

We should therefore not be surprised that domestic hunger has been outsourced by austerity minded OECD governments to corporate food charity dependent upon the wasted and surplus food of the industrial food system. Income assistance and public policy is out, food assistance and philanthropy is in, wrapped in a blanket of corporate social responsibility (or investment) and misleading promises of solidarity with the poor.

Nor should one be taken aback by Big Food's corporate capture of food banking. Certainly the food industry were invited donors and it has played a pivotal role in the global expansion of the food charity economy and food safety nets for those leading precarious lives on unliveable wages and financial assistance. Its branding and support for building secondary food markets could only be a sound business and social investment. Whether they reduce domestic hunger and food waste and ensure sustainable food access and food justice are pressing questions for civil society but ultimately for governments which must be held publicly to account.

The rise of food bank nations engages questions of compassion, solidarity, human rights and public policy: food charity or income distribution; the stigma of being fed or the ability to feed oneself and one's family with choice and human dignity. Where should collective responsibility for domestic hunger lie: the State, the private sector or civil society? More pointedly, in our affluent and 'throwaway' societies, is distributing corporately manufactured surplus or 'left-over' food to 'left-behind' people, themselves surplus to the labour market, really the best our affluent societies can do?

A necessary conversation

There is therefore an urgent need to change the conversation from one of corporate food charity and food safety nets to fair income distribution and 'joined-up' public policy informed by collective solidarity and the human right to adequate food. It is a debate about moving beyond surpluses. This reframing of the rich world hunger issue asserts that food is a basic human need, a fundamental human right and an intensely political matter. It recognizes the indivisibility of human rights; the idea that we are all 'rights holders'; the central obligation of government as the '*primary duty bearer*' for ensuring the progressive realization of the right to food including access to food for all; the vital role of international human rights instruments; and the critical role of civil society in its different national contexts for putting democratic debate and politics back into hunger and holding indifferent governments to account.

OECD food bank nations: market economy principles

With food banking now present in all 35 OECD countries debating such a change of conversation would vary widely. While the USA has been home to food banking since 1967, Sweden saw its first food bank established in 2015. Clearly OECD member states differ considerably by geography, size of population, cultural diversity, economic resources and in their constitutional arrangements and governance including significant divisions of power within and between federal, devolved and unitary states. While all are wealthy countries compared to the Global South, the histories and development of their post WWII welfare states differ markedly in universal and selective approaches to health, education and social well-being and in their policies and capacity for reducing poverty and income inequality.

Yet OECD member states have certain characteristics in common. They have all, with the exception of the USA, ratified the right to food, and, ironically, are all formal or informal members of the food bank nations' club. Most significantly, as the OECD's 50th Anniversary Vision Statement clearly states, all members share fundamental values. They 'form a community of nations committed to the values of democracy based on the rule of law and human rights, and adherence to open and transparent market-economy principles'. Furthermore the OECD's 'mission, which evolves over time, is to promote inclusive and sustainable economic growth and to raise employment and living standards' (OECD, 2017b).

Yet what practical meaning do these values and promises – 'democracy', 'rule of law', 'human rights', 'market-economy principles', 'inclusive', 'sustainable', 'employment' and 'living standards' – have for ordinary folks living 'flexible' and 'precarious' lives within or outside the labour market? Of course if you are well employed they seem purposeful and how liberal democracies should be operating. If not, for the millions dependent on charity food, they are a smokescreen for the false promises of neoliberal 'market economy principles' and 'economic growth': poverty wages, intermittent earnings, broken social safety nets and food banks. If the OECD recipe is to be more of the same changing the conversation remains a considerable challenge.

Neo-liberalism and food deprivation

As Tiina Silvasti and I wrote when summing up the findings of *First World Hunger Revisited* the majority of authors who wrote the twelve country case studies expressed 'no hope for the possibility of progressive national politics and its capability to solve the hunger issue within the context of prevailing neo-liberal economic policy' (Silvasti and Riches, 2014). Moreover this was written before Brexit and the assault of the Trump regime upon our collective psyche making more imperative the adoption of an international right to food agenda. Yet how possible might that be?

Whilst neoliberalism has brought unparalleled wealth and affluence to powerful elites in the developed world so it has ushered in deepening inequality, increasing income poverty and food deprivation for many. As George Monbiot reminds us its guiding mandate is economic growth, deregulation, lower taxes and freedom from state control, and 'freedom from tax means freedom from the distribution of wealth that lifts people out of poverty' (Monbiot, 2016).

In the process OECD member states have been transformed into food bank nations feeding 'left-over' food to 'left behind' people. It is the philosophy of *homo economicus* which understands food simply as a market commodity and not as a public, social and cultural good and a matter of democracy and human rights. Fair income distribution is not on the agenda. Neoliberalism, without a care, has failed to put food on the table for many. The indifferent State is missing in action. If progress is to be made in realizing the right to food, it is first necessary to appreciate the scale of domestic hunger in the rich world and who is benefiting and why from the establishment of corporately managed charitable food safety nets.

Persistence of domestic hunger: the problem remains

In today's Gilded Age of philanthrocapitalism 60 million people are estimated to be using food banks in high income countries, a significant marker of the scale of domestic hunger (Gentilini, 2013). Yet, as research has shown, food bank usage is a considerable underestimate of those who are hungry. The figure rises to 90 million people when judged by the FAO-FIES measure of those experiencing moderate or severe food insecurity in 33 OECD member states increasing to 140 million when including upper middle income Mexico and Turkey (Table 2.2). However these are provisional estimates only and are not based on national sources. While North America can modestly claim official and reliable counts of household food insecurity and domestic hunger, elsewhere in the OECD no one knows: robust standard measures are generally missing.

The moral imperative that people must not be allowed to go hungry goes without saying. However food banking's corporate message that charitable food banking is the link for solving the problems of hunger and food waste defies the evidence of its ineffectiveness. Rather it creates the public perception that domestic hunger is being addressed, masking the problem of poverty wages and income benefits.

Contentedly, the indifferent State carelessly looks away while passing legislation transforming corporate food waste into food donations. Leaving aside the stigma of having to beg for food to meet a basic human need as an affront to human dignity which Big Food executives or leading politicians stop to ask whether 'feeding the need' through surplus food really is an effective response to the ubiquity of food poverty. As has been said 'we go on giving and giving but the problem remains'.

The food bank benefit chain

I have argued in these pages that the corporate food bank nation is an obstacle and not a solution to domestic hunger in the rich world (or the poor world for that matter) with right to food analysis uncovering the major beneficiaries along the food bank benefit chain.

While food banks will alleviate day to day hunger for some, the advantaged are more likely the donors themselves: the corporate food sector through cost savings and tax breaks enabling the diversion of food waste from landfills to surplus food for food banks; the branding of Big Food as well as smaller scale food retailers and restaurants as good corporate citizens promoting solidarity, surely good for business; the esteem in which celebrities, the corporate entertainment, sports and media industries are held for fund raising and food donations; and the contributions, many times over, of individual volunteers doing their bit out of a sense of practical compassion enabling food banks to deliver food hampers on the front line. Hunger is thereby socially constructed and legitimized as a matter for charity, becoming de-politicized.

Yet, with the politics taken out of hunger, it is national governments which benefit the most. Outsourcing domestic hunger, the bad risks, to corporately dependent charitable food banking is accepted as an effective no cost solution, enabling the State to claim its austerity driven prescriptions are responsible public policy while repeating *ad nauseam* that work is the only way out of poverty and ever lower personal and corporate taxes will raise all boats. If not, let corporate charity and voluntary labour feed the poor.

However, food insecurity is at root a matter of income poverty requiring adequate and equitable income distribution not the redistribution of surplus food. Clearly direct food aid is required in emergency situations, for example Hurricanes Katrina, Irma and Maria in the US and Caribbean; earthquakes in Italy, Japan, Mexico and New Zealand; bush and forest fires in Australia, Canada, Portugal, Spain and the USA; and the Grenfell Tower tragedy in London, but when emergency food aid becomes a permanent daily response to persistent food poverty, it not only perpetuates the problem it claims to be solving but becomes a significant obstacle to achieving food and social justice and public policy reform.

Handing out surplus food only feeds the need for increasing volumes of corporate food waste and the expansion of ineffective food charity to cover for failing systems of publicly funded social protection and income security. The solidarity being expressed by corporate food banking as one of partnership between food as commerce and food as charity glosses over the deep irony that in the US the CEO of Feeding America is earning $650,000 per annum, and low waged food industry workers are relying on food banks (Fisher, 2017) whose army of volunteers are providing free labour. What better indicator of the wealth and income gap between the 1% and the 99%?

Inevitably food banking and its corporate reliance has become daily and normalized: there is no alternative, another repeated neoliberal mantra. Governments become passive bystanders, with corporate food charity a significant obstacle blocking progressive public policy designed to eliminate domestic hunger and reduce poverty. Yet there are always policy choices and different courses of action.

Policy choices: divergent paths

Policy-making is as much about decisions governments make as those they avoid. There is always a choice between the road most travelled and the one less so (Frost, 1916): the tried and trusted path of food charity and uncritical solidarity or the one waiting to be trodden, the path of food justice and the right to food.

Corporate food charity: the path most taken

The easy option is for the State to stand aside (sometimes lending a helping hand) by accepting food banks as a solution to widespread food poverty (and food waste). Let corporate food charity continue to feed the poor and underpin the building and strengthening of the charity food economy and focus on logistics, voluntarism and the business savvy of how to get this done. With tacit endorsement of the State it will require long-term sustainable partnerships between Big Food, corporate philanthropy, local communities and unpaid labour.

Expanding food safety nets may well be seen to express solidarity with the poor but welfare reform policies such as workfare in North America and Universal Credit in the UK will continue to push people into unstable 'flexible' poverty wage work. Food transfers will replace cash transfers and social security, enduring but threatened entitlements of the post WWII European style welfare states. For over three decades this has been the neoliberal way, an incrementally constructed, well trodden and publicly accepted way of feeding and managing the poor: institutionalizing unaccountable corporate food charity. The more secure its foundations the more difficult it will be to re-direct.

However, before continuing further along this familiar food bank route one needs to disentangle the two scandals of food waste and food poverty. The former is a symptom and downstream consequence of a dysfunctional industrial food system generating endless amounts of surplus food; the latter a result of years of austerity driven low wages, inadequate welfare benefits and unfair income distribution. Logically, it is hard to see how surplus food distribution will curtail the upstream production of food waste nor the reduction or eventual elimination of food poverty. Food banks are always running out of food. Empty shelves can only drive demand for increasing volumes of food waste. Is there a way to move beyond food surpluses?

The right to food: the path less taken

What then of the path less taken, advancing the right to food as an effective response to domestic hunger rooted in collective solidarity and social justice. This requires attracting the priority attention of the State and the development of 'joined-up' public policy: income distribution, food, housing, public health and social policy, certainly a considerable challenge in the neoliberal world.

While the right to food has made progress in certain emerging economies such as Brazil, India and South Africa by governments being held to account 'through voting, through the media: essentially through participation in public life' (Cresswell Riol, 2017), it must be acknowledged its political traction in OECD member states is weak. This much is evident judging by their lack lustre attention to the right to food when participating in the 5-year UN UPR and CESCR Periodic Reviews. For progress to occur there has to be public understanding and open debate about the right to food in principle, in practice and in action and why this matters if domestic hunger is to be eliminated and poverty reduced.

It needs public acknowledgement that food like air and water matters. All are basic human needs and fundamental to nutrition, human health and life itself. Moreover food is not only an agricultural product and an economic commodity traded, sold and bought in the market place but an environmental and public good. Food belongs to and connects us all, it is a social and cultural good. Taking bread or eating rice together is about family, friendship and community well being. It is about collective solidarity. Global and national food security lies at the centre of society and the core obligations and responsibilities of government. Food security and food access are moral, legal and political obligations of the State.

Food is too often taken for granted especially by politicians and all those who never have to think where the next meal is coming from. In market economies if you lack the financial means to shop for food in normal and customary ways like everyone else, not only will you not eat, your health will suffer and you will be excluded from society. In the rich world millions go hungry. For all the above reasons the right to food matters: it is an entitlement. The message of the right to food needs translating into everyday language. What does it mean in principle and practice in affluent OECD food bank nations?

Right to food: in principle and practice

Food access from a right to food perspective is not about charity, corporately captured or not. Neither is it about feeding hungry people by adapting North American corporatized systems of food banking for use in other wealthy OECD countries (though as *Food Bank Nations* reveals that is already well established). Nor is it about expanding secondary welfare systems and food markets and 'safety nets of the safety net' which further entrench a class of secondary food consumers (not citizens) with all the human indignity which that implies. The moral imperative to feed hungry people certainly underlines the need for nuanced understandings of food banks as potentially places of safety and referral sites to other social agencies (Lambie-Mumford, 2017). However asking for free food remains a stigmatizing experience and ineffective response to domestic hunger

Rather from the perspective of collective solidarity food access is about enabling people to feed themselves and their families with dignity and choice, essentially about people having sufficient money to shop for food like everyone else. In principle and in practice the public policy goal demands attention to income security and fair income distribution for all; and to the 'joined-up' policy ways by which that can be accomplished.

International and national law

The right to food is about the right of peoples everywhere 'to an adequate standard of living for himself and his family, including adequate food, clothing and housing, and to the continuous improvement of living conditions' and 'the fundamental right of everyone to be free from hunger' (*ICESCR*, 1966). More specifically as set out in General Comment 12, 'the right to food is realized when every man, woman and child, alone or in community … have physical and economic access at all times to adequate food or the means for its procurement' (CESCR, 1999).

As set out in these international instruments the right to food is more than rhetoric and a narrow legal injunction, it is the basis for comprehensive or what might best be termed 'joined-up' public policy and practice focused on the elimination of domestic hunger and the reduction of poverty within the context of an achievable quality of life for all. It is an expression of collective solidarity. Under international law the right to food is a legal entitlement. In terms of food access its meaning is clear and precise: that of having sufficient money to shop for food or grow your own like anyone else as fundamental to a decent standard of living.

A right to food agenda

Inescapably, this story of the rise of food bank nations, leads to the question of the meaning of the right to food in practice. It is not enough to say we are all 'rights holders' and leave it there. There is an unavoidable date with intervention requiring a right to food agenda and facing up to the challenges of implementation.

International ratification is one thing but as Olivier De Schutter, the former UN Special Rapporteur has argued, full implementation requires a constitutionally entrenched national food policy framework in the rich world's nation states (De Schutter, 2010). It means recognizing not only the indivisibility of civil and political rights and economic, social and cultural rights but also the justiciability of these rights including the right to food. A right is not a right unless it can be claimed.

Under international law the obligations to '*respect, protect and fulfil*' the right to food and ensure its progressive realization are those of the State as '*primary duty bearer*'. This requires governments not to introduce social spending cutbacks nor punitive welfare eligibility criteria which restrict people's ability to purchase food and pay the rent; to protect people from powerful 'non-State' actors including corporations or individuals who may threaten people's livelihoods (encroaching on Aboriginal lands); and to guarantee food safety for all.

Specifically, the elimination of domestic hunger, the reduction of poverty and food access for vulnerable populations demands the official monitoring of food insecurity, valid and reliable standard measures, setting targets, establishing indicators and time related benchmarks as a basis for evidence-based policy making. It requires civil servants to be educated and trained to undertake such work (and politicians who understand the reasons why).

'Joined-up' public policies

Such an agenda points to integrated or 'joined-up' public policies by connecting the policy dots between 'the four pillars of food security: availability, stability of supply, access and utilization' (FAO-VGs, 2005). The State's obligations to meet the social protection and food needs of vulnerable peoples likewise require a range of 'joined-up' food, housing, public health and social policies: fair income distribution; living wages and adequate social security and income assistance benefits indexed to the cost of living.

Interestingly the Basic Income (Guaranteed Annual Income) has recently regained political interest with new experiments currently being conducted in Finland, the Netherlands and Canada. The Basic Income may go a long way to obviating the need for food charity. However it requires cross party political support and will not be the panacea for meeting all types of human need as 'income situations can change quickly, and there would still be a need for short-term assistance' (Lightman and Lightman, 2017).

Depending on different national circumstances 'joined-up' social protection policies would also include state funded procurement of food for universal school meal programmes; public funding of street level health and social service agencies enabling them to provide the food and meals necessary for vulnerable people who even with the necessary income may be unable to purchase their food let alone cook it; the homeless; those with mental health problems, physical disabilities and drug addictions; the frail elderly; and victims of abuse. Priorities on any right to food agenda would include housing affordability; publicly funded child care; and the full funding of adult education and employment education and training.

In certain countries such as Australia, Canada, New Zealand and the United States if national reconciliation with Indigenous peoples is to mean anything, the highest priority should be partnering with those whose lands have long since been appropriated and traditional methods of food acquisition denied or threatened so as to ensure their right to food within a context of food sovereignty and democratic governance.

There is little doubt that such an inclusive right to food agenda informed by collective solidarity will not come cheaply. However, as noted, the economic and social costs of poverty and inequality in the short and long term are far more damaging and divisive. 'Joined-up' policies demand progressive taxation, particularly focused on the wealth and tax avoidance of the rich and the transnational corporate world. If such an agenda may be criticized for being grounded in the social democratic ideals for a post World War II welfare state, so be it. The fact is the neoliberal recipe dates back to Victorian times, which then as now miserably failed those waiting in the breadlines.

What then are the prospects for changing the conversation about domestic hunger from corporate charity to the right to food, as one practical step towards putting food on the table and in the process engaging wider public debates about poverty and inequality?

A hint of optimism

As austerity economics and the privatization of welfare remain alive and well in today's food bank nations perhaps looking to the indifferent State for 'joined-up' policies focused on eliminating hunger, reducing poverty and progressive taxation is wishful thinking. Yet, perhaps a hint of optimism is permitted. The fact that internationally every five years there are periodic reviews of UN member states' human rights records including economic, social and cultural rights keeps alive the idea of the right to food. The fact that the Nordic welfare states, though late comers to the food bank table, have been built on the principle of social equality informed by human rights is evidence that poverty and inequality can be success-fully addressed. The fact that the Basic Income is again on national agendas is a hint of recognition that income distribution is what counts.

Meanwhile in Canada, where we have a Prime Minister with sunny ways, the Federal Government is undertaking two separate national consultations on poverty reduction and food policy. In Mexico, the one OECD country in which the right to food is constitutionally entrenched (FAO, 2010), in 2016 it was embedded in the new constitution of Mexico City, home to more than 20 million people, devolving the State's legal obligations to realize the right to food to municipal action and accountability (Castrejon-Violante, 2017c; see also FAO, 2010).

A word of caution. The persistence of widespread food poverty and indifferent states' dependence on corporate food charity's perceived effectiveness suggests such initiatives may be little more than tinkering, a continuation of the neoliberal status quo and 'uncritical' solidarity with the poor. Economic and political uncertainties created by Brexit and the Trump regime seem to rule out the possibility of imminent progressive change. Understandably poverty reduction and fair income distribution informed by the right to food is inspired more by hope than any beckoning commitment by governments to food and social justice.

Yet as Rebecca Solnit's book *Hope in the Dark. Untold Histories Wild Possibilities* reveals, progressive change is always possible. It is a question of expectations. As she writes hope 'is not the belief that everything was, is or will be fine. The evidence is all around us of tremendous suffering and tremendous destruction'. Rather the hope Solnit is 'interested in is about broad perspectives with specific possibilities, ones that invite or demand that we act. It's also not a sunny everything-is-getting-better narrative (*pace* Justin Trudeau – author comment), though it may be counter to the everything-is-getting-worse narrative. You could call it an account of complexities and uncertainties, with openings' (Solnit, 2016).

In terms of the crisis of domestic hunger in the rich world it is then about finding the openings and generating public support and political will for the right to food and social justice agenda informed by collective solidarity. It is about moving beyond corporate food charity, putting the politics back into hunger and gaining the priority attention of indifferent governments. It is about the choices we make.

Challenging propositions for public debate

Taking the path of collective solidarity and the right to food requires not only a stiff tonic of hope and resolve but also a huge shot of political will. The failure of politicians, including those with progressive leanings to summon the political courage to act upon the ineffectiveness and hidden functions of corporate food charity is a serious obstacle preventing the elimination of domestic hunger, the development of national actions plans and governments accepting public accountability.

The austerity driven indifferent State, even in tolerant and fair-minded Canada, always chasing lower taxes and ignoring broken welfare systems and the need for living wages, adequate income benefits and joined-up food, public health and social policy presents a formidable challenge. A protest for years has been a sticker, courtesy of the Toronto-based Centre for Social Justice, on the front door of economist and social policy emeritus professor and friend, Ernie Lightman, which reads:

MORE POVERTY. YOUR TAX CUTS AT WORK

a key reason why questions need asking to stimulate public debate and generate political will for a right to food agenda. It needs civil society with a right to food bite as the catalyst for changing the conversation.

It is not possible nor even advisable to advance a prescriptive right to food agenda for change within 35 OECD food bank nation states. There does however need to be public recognition that corporate food charity is an ineffective and stigmatizing response to domestic hunger. It may be a neoliberal dream come true for philanthrocapitalism but is a daily nightmare for those dependent on wasted and surplus food. As Yvonne Kelly of Freedom 90, the union of food bank volunteers in Ontario, has said 'the hunger problem is not going away. What we are doing is not solving it' (YK, phone 4.10.17). There is a need to change the conversation to the right to food and engage public debate about a number of challenging propositions:

- acknowledging that corporately dependent charitable food banking far from being the best affluent food secure OECD member states can do is an obstacle to eliminating domestic hunger and reducing poverty in the rich world;
- moving 'beyond surpluses' means separating the scandals of food waste and food poverty from each other and rejecting the idea that the food bank twinning of these two issues is a 'win-win' solution for both. The global and dysfunctional industrial food system sending corporate food waste to patch up broken social safety nets and feed hungry people is neoliberal doublespeak perpetuating both problems, never a solution;
- recognizing that food poverty in wealthy food secure nations with cheap food policies is primarily caused by income poverty and inequality and not a lack of food supply; and the need for a national and routinely collected standard measure of food insecurity to inform evidence-based policy directed at the elimination of domestic hunger;
- rejecting 'uncritical' solidarity and adopting the perspective of collective solidarity demanding to know who benefits and why from the privatization,

outsourcing and downloading of domestic hunger to corporate food charity reliant on the unpaid free labour of food bank volunteers;

- understanding that access to food for those leading precarious lives means connecting the public policy dots between stable work, living wages, adequate benefits and progressive taxation and between pro-poor food, housing, public health and social policy recognizing agricultural and environmental imperatives;
- partnering with Indigenous peoples in countries such as Australia, Canada, New Zealand and the USA whose lands have long since been appropriated and traditional methods of food acquisition denied or threatened to advance national reconciliation by according the highest priority to their right to food within the context of food sovereignty;
- resisting philanthrocapitalism, the further Americanization of rich world welfare states and the false promises of corporate food charity establishing 'food safety nets' for 'the safety net' which undermine long-established European systems of social security informed by income distribution and human rights.
- making clear that the right to food is not about charity, nor empty rhetoric, nor a legal quick fix, but under international law holds the State morally, legally and politically accountable as the '*primary duty bearer*' for achieving food security for all while respecting the right of everyone to feed themselves and their families with choice and human dignity;
- challenging the Indifferent State to learn from the advocacy of NGOs such as FIAN International, Food Secure Canada, the UK's Food Research Collaboration and Nourish Scotland, and cross-party Scottish politicians who, shamed by food banking, are putting politics and human rights back into the hunger debate directed at food and public policy for all.

These are not easy conversations. Particularly in times of austerity, attracting public support for the idea that corporate food charity is an ineffective response to domestic hunger and that food banking is not in solidarity with the poor will likely prove the most difficult. Food banking is after all an expression of practical compassion, about volunteers 'doing their bit' to help others. Yet Solnit's 'complexities and uncertainties' provide an 'opening' for a long overdue public conversation about corporate food charity: food bank volunteers are doing their bit, but how about governments stepping up to the plate.

This moves the discussion from charity to human rights and the role of the State. Under international law it understands addressing hunger and access to food as moral, legal and political obligations which should not be left to the residual and ineffective efforts of corporate food charity, but raised as matters of collective solidarity requiring public debate and the active participation and leadership of civil society.

Putting rights and politics back into hunger

As Samuel Moyn, professor of law and history has argued, contemporary human rights need to be more than 'a set of global moral principles and sentiments', because presented as such they do not 'intervene in power politics', consequently seeming 'to make little practical difference, amounting to an ornament on a tragic world that they do not transform' (Moyn, 2014).

Moyn argues that 'a politics of human rights must start with international human rights ideas and movements as they currently exist, and radicalize them from there'. A politics of human rights must become transformational, mobilized at the grass roots and transcend judges who cannot be relied upon unless they are allied with grass roots political movements. Invoking formal rights is not enough.

Rather he sees human rights being 'contingent tools of social organisation', in other words a necessary strategy for any progressive food policy organization or CSO. Specifically, a politics of human rights must move away 'from framing norms individualistically and should cease to privilege political and civil liberties'. He further insists that 'for the sake of the common good or social solidarity, that the real conditions for the enjoyment of any rights are to be sought not simply in the possession of personal security but in the entitlement to economic welfare' (Moyn, 2014, p.62).

Civil society with a right to food bite

As the corporate capture of charitable food banking has become an institutionalized and publicly accepted response to domestic hunger, Moyn's human rights analysis has particular resonance. Such rights talk challenges civil society and its NGOs to become the catalyst or social animateur for changing the public conversation from food as charity to one of rights and justice. It calls for civil society action which informs public and professional education and debate and energizes and empowers the 'collaborative' and 'adversarial' roles of public participation (see Drèze and Sen, 1989) necessary for generating political will. It is about civil society with a relentless human rights appetite, and bite, for actively advancing the moral, legal and obligations of the right to food agenda.

Yes, NGOs, CSOs, charities, think tanks, academia – should as a matter of principle and pragmatism adopt collaborative approaches in seeking constructive dialogue with duly elected governments. However, there may be little choice but to be adversarial and partisan. The right to food offers a counter-narrative standing in clear opposition to more than three decades of free market neoliberalism, that 'pernicious economic doctrine' (Solnit, 2016); to the shameful indifference of governments neglecting the basic needs of vulnerable citizens; and to the institutionalization of corporate food charity symbolizing the scandal of 21st-century breadlines in our affluent and throwaway societies.

Fortunately acting on the right to food is not a civil society process which has to be invented for the rich world. This much is evident for example from the work of FIAN International in Europe (e.g., Austria, Belgium, Austria and Germany, Norway), Food Secure Canada and Nourish Scotland in *monitoring* their countries' domestic compliance through the UN periodic review processes with the right to food, and through the more recent activities of the GNRtFN in the USA. Such actions reflect substantive HRBA work undertaken in OECD countries by a range of CSOs and NGOs concerned with food policy, food ethics, food justice, food sovereignty; nutrition and public health; social policy and social justice; anti-hunger and poverty reduction; community food security; civil liberties and human rights.

CSOs operate at the grass-roots as food policy councils; professional associations; research centres; food action networks; community organizations and neighbourhood houses and by building coalitions locally, nationally, internationally and at the level of the United Nations. In addition to monitoring activities their work includes researching, advocating, campaigning, coalition building and public education from food, public health and social policy perspectives in formed by human rights.

Teaching, research, public education and coalition building undertaken by units such as the Centre for Food Policy (CFC) at City University in London with its extensive global and UK academic reach; PROOF, the interdisciplinary research team at the University of Toronto investigating household food insecurity in Canada (connecting researchers across the country and in the USA); and the Centre for Hunger-Free Communities in the Dornsife School of Public Health at Drexel University in Philadelphia shape the food policy and right to food discourse (CHFC, 2017). These are just three examples of such academic activity. They provide authoritative evidence-based studies and critical analysis of key food system and food poverty issues, and are training the next generation of food policy researchers and activists.

Coalition building and advocacy: Such organizations not only link food interested academics working across different disciplines but play a critical role in *coalition building* and *advocacy*. The CFC, for example, launched the Food Research Collaboration in the UK in 2014 which brought together CSOs and academics 'to produce, share and use the evidence based knowledge needed to influence and improve UK food policy' (FRC, 2016). FRC partnered with Sustain, the Food Foundation, Oxfam and the Sociology Department of Oxford University in the 2016 workshop advocating that the UK should adopt a standard measure of food insecurity (FF, 2017), an essential component of a right to food and public policy agenda.

Campaigning and advocacy: significantly national food policy CSOs which bring to public attention the adverse impact of Big Food's control of the global food system are crucial players for combatting the corporate capture of food charity. Whilst they may not explicitly focus on the right to food, its language has begun to inform the goals and strategies of some food justice organizations. These obviously include the FIAN national sections in Europe, the UK's Food Ethics Council, the Scottish Food Coalition, Food Secure Canada, Canada Without Poverty, and the newly formed Right to Food Coalition in Australia. WhyHunger in the USA is also discussing the right to food.

Indeed the first of the SFC's four core principles – People Matter – unequivocally states: 'everyone has the right to sufficient, safe, nutritious, and culturally appropriate food obtained in ways free from stigma or status, now and into the future' (SFC, 2017). Nourish Scotland is generating political will through its rights-based

'Menu for Change' and its advocacy leading to proposed 'Good Food Nation' legislation in the Scottish Parliament. The first of Food Secure Canada's 'Five Big Ideas' is the right to food. In these ways commitments are made to address domestic hunger through *campaigning* and *advocacy* by connecting the dots within national food policy framework legislation.

Advocacy and learning at the local level may have a long reach. The rights-based approaches of Witnesses to Hunger at Drexel University in Philadelphia give voice to those living on poverty budgets, 'the real experts', in monitoring and researching solutions to hunger and economic insecurity, and putting food and health promotion ideas into practice at the local level but have also taken their *advocacy* to the highest level – the US Congress and the White House (CHFC, 2017). The PotLuck Café in Vancouver's Downtown Eastside provides training and paid positions for those with difficulties finding traditional employment and assists with finding jobs elsewhere. The BC Poverty Reduction Coalition as with Put Food in the Budget in Ontario campaigns and advocates for raising the social assistance rates while the Basic Income Group in Ontario is pressings its case. Similarly Canada Without Poverty advocates through its Dignity for All campaign on behalf of the right to food, housing and adequate incomes.

Such national and local expressions of civil society educating and campaigning with a right to food bite is based on a shared understanding that food poverty is primarily caused by lack of money, and food access must be assured by adequate wages and social security benefits to meet the costs of living and by progressive income distribution. The focus is holding the State to account and putting the politics back into hunger. Food banking dependent on wasted and surplus food supplied by the corporate food sector is not on the menu because it is experienced and viewed as ineffective and stigmatizing, an affront to personal dignity and human rights.

We are doing our bit. Where is the State?

Yet having argued that food banks are part of the problem not the solution, is there a role for corporate food charity in advancing the right to food? In my view their contradictory functions constitute a considerable obstacle. Surely food banks have proved to be durable and are publicly perceived to be necessary stop gaps. With a strong sense of practical compassion they are feeding hungry people, but the cloak of non-political corporate social responsibility sits uncomfortably with the politics of food and social justice.

The business of charitable food banking, institutionally dependent on Big Food brands, must always focus on stocking shelves and the logistical imperative of building more efficient surplus food distribution hubs. While the websites of national food bank organizations make links between food waste, hunger and poverty, and some such as the UK's Trussell Trust may publicly advocate necessary social policy reforms, their primary task has to be increasing their supplies of surplus or donated food, even seeking tax exemptions from the State. Their corporate backers however are unlikely to press the case for living wages, increasing social security benefits and higher rates of corporate taxation.

If there is a source of advocacy within the food bank movement it needs the volunteers to step up to the plate. After all, the whole edifice of Foodbanks Inc depends upon their tireless giving and unpaid labour. Unchanged today is the concern expressed by Eldon Anderson, the first chair of the Regina Food Bank in Canada in 1983, that handing out free food to hungry people, while driven by a moral imperative should be unacceptable in today's wealthy and food secure societies. Indeed it has been called stigmatizing and shameful. Such ambivalence is an open invitation to change the conversation building on the pioneering work of Freedom 90, the union of Ontario food bank volunteers campaigning to make food banks obsolete.

Even in the USA, as Andy Fisher writes, food bank staff concerned about 'the collateral damage being done to the spirit of the poor, and the poor quality of the food they distribute', want to do things differently including using different indicators of food bank success than measuring pounds and people. Higher wages, and reduced income inequality are on the US anti-hunger agenda (Fisher, 2017, p.265). As for generating political will, one strategy for ambivalent and willing food bank volunteers would be each time they give a food parcel to a person in need, it should be accompanied by a tweet, email or postcard to all elected politicians in their respective OECD member states, saying 'We are doing our bit, where is the State? Act on your obligations under international law'. It would need to be relentless messaging until domestic hunger is eliminated.

A positive sign would be the stirring of a human rights and political debate about the use of 'left-over' food in affluent societies as a front line response to feeding millions of 'left-behind' people. It must be a debate about the moral vacuum at the centre of free-market neoliberalism and the indifference of austerity minded governments responsible for shrunken welfare states and punitive welfare policies. It must be about resisting and moving beyond the adoption of ineffective corporate style US food banking in OECD member states historically better served by income security and universal social programs.

A final reflection

A guiding motif for this book has been French historian Fernand Braudel's comment, written shortly before he passed away in 1985 that 'today's society, unlike yesterday's, is capable of feeding its poor. To do otherwise is an error of government' (Braudel, 1985). Clearly, the governments of the wealthy OECD food bank nations were as indifferent then as they remain today. Given that in the UK there is at least a political debate about food poverty and food banks I would like to end with a prescient and compelling observation from eminent UK philosopher Onora O'Neill, likewise written in 1985 at the time when food banking was beginning to take hold in the rich world, that 'a serious commitment to charity and beneficent action requires commitment to material justice and so to political change. Practical reasoning about hunger has an audience only when it reaches those with power to bring about that change' (O'Neill, 1986).

Engaging the priority attention of affluent nation states to act on their moral, legal and political obligations to implement the right to food requires the relentless participation and active political bite of civil society. For all those standing in the breadlines, the focus must be on food access, 'joined-up' food policy and social justice: living wages, adequate social security benefits, fair income distribution and progressive taxation. Lacking this the shelf life of the OECD food bank nations and the charity food economy will become further entrenched, a permanent indicator of dysfunctional food systems, broken social safety nets and the erosion of social rights, incapable of eliminating widespread hunger in the rich world.

There is a choice: public policy informed by the right to food.

REFERENCES

AD (2015) *Sustainable Retailing*, Ahold Delhaize https://www.aholddelhaize.com/media/1934/ahold-responsible-retailing-report-2015.pdf

A2H (2006) *Annual Report*, America's Second Harvest

Adzakpa, H. (2016) *The Right to Adequate Food. A comparative analysis of EU member states' reporting on and implementation of the Right to Food under the International Covenant on Economic, Social and Cultural Rights*. Research Report, Nourish Scotland 30 July http://www.nourishscotland.org/campaigns/right-to-food/nourish-report-to-un-cescr/

Alcock, P. and Craig, C. eds (2009) *International Social Policy. Welfare Regimes in the Developed World*. 2nd edition. Palgrave Macmillan

Anderson, M.D. (2013) Working toward the right to food in the USA. Ch. 10d in *Alternatives and Resistance to Policies that Generate Hunger*, Right to Food and Nutrition Watch. Impressum http://www.fian.org/fileadmin/media/publications_2015/Watch_2013_eng_WEB_final.pdf

Anderson, S.A. ed. (1990) Core indicators of nutritional state for difficult-to-sample populations. Life Sciences Research Office, *The Journal of Nutrition* 120, 1557–1600

APN (2016) *Food Loss and Waste in the Agro-Food Chain*, Organisation for Economic Co-operation and Development https://www.oecd.org/tad/policynotes/food-loss-waste-agro-food-chain.pdf

APPG (2014) *Feeding Britain. A strategy for zero hunger in England, Wales, Scotland and Northern Ireland*. Report of the All-Party Parliamentary Inquiry into Hunger in the United Kingdom.

APPG (2015) *Feeding Britain. Six Months On, A Progress Report on All-Party Parliamentary Inquiry into Hunger in the United Kingdom* https://feedingbritain.files.wordpress.com/2015/06/feeding-britain-six-months-on.pdf

Arbour, L. (2005) 'Freedom from want' – from charity to entitlement. LaFontaine-Baldwin Lecture, Quebec City

Arbour, L. (2008) Address by Ms. Louise Arbour, UN High Commissioner for Human Rights on the 7th Special Session of the Human Rights Council, Geneva 22 May https://reliefweb.int/report/world/address-ms-louise-arbour-un-high-commissioner-human-rights-occasion-7th-special-session

AUS (2016) *Australia-Fifth Periodic Report due in 2014*, UN Economic and Social Council, E/C.12/AUS/5 1 February http://tbinternet.ohchr.org/_layouts/treatybodyexternal/Download.aspx?symbolno=E%2fC.12%2fAUS%2f5&Lang=en

Ballard, T.J., Kepple A.W. and Cafiero, C. (2013) *The food insecurity experience scale: development of a global standard for monitoring hunger worldwide.* Technical Paper. Rome, FAO www.fao.org/fileadmin/templates/ess/VOH/FIES_Technical_Paper_v1.1.pdf

Barnard, C. (2014) Solidarity and the Commission's 'Renewed Social Agenda'. Ch. 4 in Roos, M. and Borgmann-Prebil, Y. *Promoting Social Solidarity in the European Union.* OUP

Bauman, Z. (2004) *Wasted Lives. Modernity and its Outcasts.* Polity Press

Bazerghi, C., McKay, F.H. and Dunn, M. (2016) The role of food banks in addressing food insecurity: a systematic review. *Journal of Community Health* 41, 732–740 Springer 4 January

BBC (2016) Can Sweden tackle the throwaway society? http://www.bbc.com/news/business-37419042

BBC (2017) Hunger Prevention, Feeding America/AD Council 11 August https://ads.psacentral.org/ads/HungerPrevention/LiveAnnouncerCopy/HUN_August13_Eng_LAC.pdf

Bio/Deloitte (2014) *Comparative Study on EU Member States' legislation and practices on food donation. Executive Summary.* Bio by Deloitte. Brussels: EESC, June http://www.worldcat.org/title/comparative-study-on-eu-member-states-legislations-and-practices-on-food-donations-final-report/oclc/903501464

Berg, J. (2008) *All You Can Eat. How Hungry is America?* Seven Stories Press

Bishop, M. and Green, M. (2008) *Philanthrocapitalism: How the Rich Can Save the World*, Bloomsbury

Bloom, J. (2010) *American Wasteland.* Da Capo Press

BMGF (2014) Bill & Melinda Gates Foundation https://en.wikipedia.org/wiki/Bill_&_Melinda_Gates_Foundation

Booth, S. (2014) Food banks in Australia. Discouraging the right to food. Ch, 2 in Riches, G. and Silvasti, T. eds *First World Hunger Revisited.* Palgrave Macmillan

Boswell, L. (2015) FareShare encourages a level playing field for food disposal, FareShare http://www.fareshare.org.uk/fareshare-encourages-level-playing-field-for-food-disposal/

Braudel, F. (1985) The New History, *World Press Review* www.worldpress.org

BRC (2016) *The Retail Industry's Contribution to Reducing Food Waste, British Retail Council, London*, Autumn http://brc.org.uk/media/105811/10105-brc-food-waste-report-final.pdf

Butler, P. (2014) 'Need for food banks is caused by welfare cuts, research shows'. *The Guardian*, 8 April

Butler, P. (2017) 'Foreword' in Lambie-Mumford, H. *Hungry Britain. The Rise of Food Charity.* Policy Press.

BW (2009) Key Hunger Terms, Bread for the World Institute, Washington, DC http://www.bread.org/learn/hunger-basics/key-hunger-terms.html

BW (2014) Fact Sheet: Churches and Hunger, Bread for the World On-line http://www.bread.org/sites/default/files/downloads/2014_churches_hunger_fact_sheet.pdf

CAN (2013) *Sixth Report of Canada on the United Nations International Covenant on Economic, Social and Cultural Rights*, January 2005 – December, 2009. Minister of Public Works and Government Services Canada, Ottawa E/C.17/CAN/6

Caplan, P. (2016) Big society or broken society? Food banks in the UK. *Anthropology Today*, 32(1), 5–9 February

Caraher, M. and Coveney, J. (eds) (2016) *Food Poverty and Insecurity: International Food Inequalities.* Springer

Caraher, M. and Furey, S. (2017) *Is it appropriate to use surplus food to feed people in hunger? Short-term Band-Aid to more deep rooted problems of poverty*, Food Research Collaboration

Policy Brief 26 January http://foodresearch.org.uk/wp-content/uploads/2017/01/Final-Using-food-surplus-hunger-FRC-briefing-paper-24-01-17.pd

Cargill (2015) Enhancing Food Banks to End Hunger, 11 May http://minnesota.cbslocal.com/2015/05/11/enhancing-food-banks-to-fight-hunger/

CaribFlame (2016) Millions of Tons of Food are Wasted in Mexico, CaribFlame 15 November http://www.caribflame.com/2016/11/millions-of-tons-of-food-are-wasted-in-mexico/

Caritas Report (2014) *The European Crisis and its Human Cost, Caritas-Europa, Belgium* http://tinyurl.com/puz.3r55

Carolan, M. (2011) *The Real Cost of Cheap Food*, Earthscan from Routledge

Castrejon-Violante, L. (2017a) Food Security in Mexico, National Monitoring. Unpublished paper, Interdisciplinary Doctoral Studies Program, UBC

Castrejon Violante, L. (2017b) Interview with Federico González Celaya, President, Banco de Alimentos de México (BAMX) 8 March

Castrejon-Violante, L. (2017c) The Right to Food in the new Mexico City Constitution, unpublished paper, Inter Disciplinary Studies Programme, University of British Columbia

CB (2017) Five Canadians are as rich as 30% of the population, not two, Canadian Business http://www.canadianbusiness.com/lists-and-rankings/richest-people/oxfam-canada-inequality-richest-people/

CBC (1985) The Poverty Line. The Fifth Estate. Canadian Broadcasting Corporation, 26 February

CBC (2010) Sarah McLachlan makes food bank appeal. CBC News, December 15, 2010, www.cbc.ca/beta/news/canada/british-columbia/sarah-mclachlan-makes-food-bank-app eal-1.892056

CBC (2012) UN official sparks debate over Canadian food security. CBC News Toronto, May 16, 2012 http://www.cbc.ca/news/politics/un-official-sparks-debate-over-canadia n-food-security-1.1130281

CBC (2015) CBC Open House & Food Bank Day Incentive Prizes, 7 December www.cbc.ca/news/canada/british-columbia/events/cbc-food-bank-day-incentive-prizes-2015–2011.3313057

CCHS (2016) *The Household Food Security Survey Module (HFSM)*, Health Canada, Ottawa http://www.hc-sc.gc.ca/fn-an/surveill/nutrition/commun/insecurit/hfssm-mesam-eng.php

CDA (2015) *Le Droit à une alimentation adéquate en Belgique.* Coalition pour le Droit a l'Alimentation, FIAN Belgium, Bruxelles June

CEFB (2017) *Charter of European Food Banks, Annex A, Federation of European Food Banks* http://www.eurofoodbank.org/about-us

CESCR (1999) *General Comment No 12*, Committee on Economic, Social and Cultural Rights E/C.12/1999/5 12 May http://www.fao.org/fileadmin/templates/righttofood/documents/RTF_publications/EN/General_Comment_12_EN.pdf

CESCR (2017) *Committee on Economic, Social and Cultural Rights* http://www.ohchr.org/EN/HRBodies/CESCR/Pages/CESCRIntro.aspx

CF (2013) International Solidarity, Press release, Carrefour Foundation 15, November http://www.carrefour.com/news-releases/international-operation-collect-food-products

CFREU (2000) *Charter of Fundamental Rights of the European Union /C 364/01* http://www.europarl.europa.eu/charter/pdf/text_en.pdf

CH12,3 (2017) *Champions 12.3, UN Social Development Goals, Post-2015 Development Agenda* https://champions123.org/about/; https://champions123.org/target-12-3/

CHFC (2017) Centre for Hunger-Free Communities, Dornsife School of Public Health, Drexel University, Philadelphia USA http://www.centerforhungerfreecommunities.org/our-projects/witnesses-hunger

Chrisafis, A. (2016) French law forbids food waste by supermarkets. *The Guardian* 4 February https://www.theguardian.com/world/2016/feb/04/french-law-forbids-food-waste-by-supermarkets

CONEVAL (2015) *Comunicado de prensa No. 005, CONEVAL informa los resultados de la medición de la pobreza 2014*

Cresswell Riol, K.S.E. (2017) *The Right to Food Guidelines, democracy and citizen participation: country case studies.* Routledge Studies in Food, Society and the Environment. Routledge

DEE (2017) Working together to reduce food waste in Australia, Department of Environment and Energy, Australian Government http://www.environment.gov.au/protection/national-waste-policy/food-waste

Deloitte (2013) *The Food Value Chain. A challenge for the next century.* https://www2.deloitte.com/content/dam/Deloitte/global/Documents/Consumer-Business/dttl_cb_Food%20Value%20Chain_Global%20POV.pdf

De Schutter, O. (2008) *Analysis of the World Food Crisis by the UN Special Rapporteur on the Right to Food.* UN Human Rights Council, 2 May

De Schutter, O. (2010) *Countries tackling hunger with a right to food approach.* Briefing Note No. 1 UN Special Rapporteur on the Right to Food OHCHR http://www2.ohchr.org/english/issues/food/docs/Briefing_Note_01_May_2010_EN.pdf

De Schutter, O. (2011) *Reports of the Special Rapporteur on the Right to Food.* Economic and Social Council: Louvain/Geneva http://www.srfood.orgUN

De Schutter, O. (2013). Food banks can only plug the holes in social safety nets. *The Guardian,* 27 February https://www.theguardian.com/commentisfree/2013/feb/27/food-banks-social-safety-nets

De Schutter, O. (2017) The Coming Revolution: The transformative power of the right to food, Department of Law, London School of Economics, 12 January http://www.lse.ac.uk/Events/2017/01/20170112t1830vTWR1G01/The-Coming-Food-Revolution

De Schutter, O. (2017) *Country Missions.* Official website www.srfood.org/en/country-mission

De Schweinitz, K. (1961) *England's Road to Social Security.* A.S. Barnes & Company, Inc.

Dey, K. and Humphries, M. (2015) Recounting food banking: a paradox of counter-productive growth. *Third Sector Review,* 21, 2: 129–147

DFI/Oxfam (2017) *The Commitment to Reducing Inequality Index,* Research Report, Development Finance International and Oxfam. https://www.oxfam.org/en/research/commitment-reducing-inequality-index

Dowler, E. (2003) Food and Poverty: Insights from the 'North'. *Development Policy Review,* 21(5–6): 569–580

Dowler, E. (2014) Food banks and food justice in 'Austerity' Britain. Ch. 12 in Riches, G. and Silvasti, T. *First World Hunger Revisited.* Palgrave Macmillan

Dowler, E. and Jones Finer, C. eds (2003) *The Welfare of Food. Rights and Responsibilities in a Changing World.* Blackwell.

Dowler, E. and O'Connor, D. (2012) Rights-based Approaches to Addressing Food Poverty and Food Insecurity in Ireland and the United Kingdom. *Social Sciences and Medicine* 74(1), 44–51

Drèze, J. and Sen, A. (1989) *Hunger and Public Action.* Clarendon Press

EC (2010) *Preparatory Study on Food Waste across EU 2,* Technical Report – 054 European Commission

E/C.12 (1993) *Concluding observations of the Committee on Economic, Social and Cultural Rights: Canada,* UN Economic and Social Council E/C.12/1993/5 3 June

E/C.12 (2016) *Concluding observations on the sixth periodic report of Canada,* UN Committee on Economic, Social and Cultural Rights E/C.122/CAN/CO/6 23 March

EComm (2016) *Euro Commerce* http://www.eurocommerce.eu/retail-and-wholesale-in-europe.aspx

EC-MDP (2008) *Food Distribution Programme for Europe's Most Deprived People Results of the online consultation.* European Commission Directorate-General for Agriculture and Rural Development https://ec.europa.eu/agriculture/sites/agriculture/files/consultations/most-deprived/results_en.pdf

EFAP (2017) Emergency Food Assistance and Soup Kitchen-Food Bank Program https://en.wikipedia.org/wiki/Emergency_Food_Assistance_and_Soup_Kitchen-Food_Bank_Program retrieved 10. 2. 17

Eide, A. (1998) The human right to adequate food and freedom from hunger. Ch 1 in *The Right to Food in Theory and Practice*, FAO Information Division, Office of Director-General

Ekwall, B. (2008) Creating Capacity through Curriculum Development – The Right to Food. Powerpoint, FAO Right to Food Unit, University Hunger Summit, Washington, 29 February

Elver, H. (2017) Written text of video message. Hilal Elver, SRRF. The Right to Adequate Food Event, 24 January http://www.ohchr.org/Documents/Issues/Food/Event24Jan2017.pdf

EPA (2017) United States 2030 Food Loss and Waste Reduction Goal, Environmental Protection Agency https://www.epa.gov/sustainable-management-food/united-states-2030-food-loss-and-waste-reduction-goal

EPSR (2017) *European Pillar of Social Rights*, European Commission https://ec.europa.eu/commission/priorities/deeper-and-fairer-economic-and-monetary-union/european-pillar-social-rights_en

Escajedo San-Epifanio, L. and De Renobales Scheifler, M. eds (2015) *Envisioning a future without food waste and food poverty. Societal challenges.* Wageningen Academic Publishers

ESCR-Net (2015) see also the NGO Coalition of OPESCR Rights https://www.escr-net.org/petitions/2015/join-ngo-coalition-op-icescr http://ec.europa.eu/environment/eussd/pdf/bio_foodwaste_report.pdf

ERS-USDA (2015) *Household Food Security in the United States in 2015*, ERR-215 www.ers.usda.gove/webdocs/publications/err215/err-215.pdf

Eurostat (2016a) *Smarter, greener, more inclusive? Indicators to support the Europe 2020 Strategy*, Luxembourg: Publication of the European Union, 2016 Edition http://ec.europa.eu/eurostat/statisticsexplained/index.php/Smarter,_greener,_more_inclusive_-_indicators_to_support_the_Europe_2020_strategy

Eurostat (2016b) News Release 2016 7 October

FA (2014) *Hunger in America, National Report for Feeding America* http://help.feedingamerica.org/HungerInAmerica/hunger-in-america-2014-full-report.pdf

FA (2015) *Annual Report 2015*, Feeding America http://www.feedingamerica.org/about-us/about-feeding-america/annual-report/2015-feeding-america-annual-report.pdf

FA (2016) *Nourishing Healthy Futures, Annual Report*, Feeding America http://www.feedingamerica.org/about-us/about-feeding-america/annual-report/2016-feeding-america-annual-report.pdf

FA (2017a) What is a food bank? Our Network, Feeding America http://www.feedingamerica.org/our-work/food-bank-network.html

FA (2017b) Our History, Feeding America http://www.feedingamerica.org/about-us/about-feeding-america/our-history/

FA (2017c) Tax Benefits for your Company, Feeding America http://www.feedingamerica.org/ways-to-give/give-food/become-a-product-partner/tax-benefits-for-your-company.html

FA (2017d) Look to the Stars: Celebrity Supporters. Feeding America https://www.look tothestars.org/charity/feeding-america

FA (2017e) Entertainment Council, Feeding America www.feedingamerica.org

FA (2017f) Food Waste in America, Feeding America http://www.feedingamerica.org/a bout-us/how-we-work/securing-meals/reducing-food-waste.html

Fabian Society (2015) *Hungry for Change*. Final Report of the Fabian Commission on Food and Poverty, London info@fabian-society.org.uk

FAO (2002) *Declaration of the World Food Summit: Five Years Later* http://www.org/3/ Y7106E09.htm

FAO (2008) What the Right to Food is Not, Right to Food Unit, Food and Agricultural Organisation. www.fao.org/righttofood

FAO (2009) *Declaration of the World Food Summit on Food Security, World Summit on Food Security, Rome*, 16–18 November

FAO (2010) *Right to Food in the Cities: focus on Mexico legislation* Right to Food Team, http:// www.fao.org/righttofood/publications/publications-detail/en/c/49559/

FAO (2014) *SAVE FOOD: Global Initiative on Food Loss and Waste Reduction*, UN Food and Agricultural Organization http://www.fao.org/save-food/resources/keyfindings/en/

FAO (2015a) *Food Security Statistics*. UN Food and Agricultural Organization http://www.fao. org/economic/ess/ess-fs/en/

FAO (2015b) *SAVE FOOD Global Initiative on Food Loss and Waste Reduction*, 2015 http:// www.fao.org/3/a-i4068e.pdf http://www.fao.org/save-food/en/

FAO (2016) *Food Balance Sheets* http://www.fao.org/faostat/en/#data/FBS

FAO (2017a) *The Right to Food around the Globe*, UN FAO http://www.fao/right-to-food-around-the-globe/en/

FAO (2017b) *The right to food in national constitutions*. FAO Corporate Document Repository http://www.fao.org/docrep/w9990e/w9990e12.htm

FAO, IFAD, UNICEF, WFP and WHO (2017) *The State of Food Security and Nutrition in the World 2017. Building resilience for peace and food security*. Rome, FAO.

FAO-ESP (2017) *Social Policies and Rural Institutions*, UNFAO http://www.fao.org/economic/ social-policies-rural-institutions/en/

FAO-FIES (2014) *Global Food Insecurity Experience Scale*, FAO http://www.fao.org/fileadm in/templates/ess/voh/FIES_062014.pdf

FAO-FISW (2015) *The State of Food Insecurity in the World 2015. Meeting the 2015 international hunger targets: taking stock of uneven progress.*

FAO, IFAD and WFP (2015) *The State of Food Insecurity in the World 2015. Meeting the 2015 international hunger targets: taking stock of uneven progress*. Rome, FAO.

FAO, IFAD and WFP FAO-RTFT (2017) *Right to Food Team*, Food and Agricultural organization, Rome http://www.fao.org/righttofood/knowledge-centre/en/

FAO-VGs (2005) *Voluntary Guidelines on the Right to Food*, United Nations Food and Agricultural Organization http://www.fao.org/3/a-y7937e.pdf

FAO-VOH (2016) *Voices of the Hungry: Methods for estimating comparable rates of food insecurity experienced by adults throughout the world*. Rome, FAO. http://www.fao.org/3/ c-i4830e.pdf

FareShare (2017) FareShare. *Fighting Hunger. Tackling food Waste*. UK http://www.fareshare. org.uk/ http://www.fareshare.org.uk/giving-food/ http://www.fareshare.org.uk/who-we-work-with/ http://www.fareshare.org.uk/our-history/

Farb, P. and Armelagos, G. (1980) *Consuming Passions: The Anthropology of Eating*. Houghton Mifflin Company

FB (2014) Food Bites: Graham Riches, Centre for Food Policy, City University https:// www.youtube.com/watch?v=nEJabSPlJQ4

FBAsia (2012) 2HA Chair Charles McJilton Visits the National Food bank in South Korea. http://foodbank.asia

FBC (2012a) *HungerCount 2012* Food Banks Canada https://www.foodbankscanada.ca/getmedia/3b946e67-fbe2-490e-90dc-4a313dfb97e5/HungerCount2012.pdf.aspx

FBC (2012b) *Stimulating Canada's Charitable Sector. A Tax Incentive Plan for Charitable Food Donations.* Food Banks Canada. Toronto. January https://www.foodbankscanada.ca/getmedia/3940f0c5-9363-4512-9852-b5ecf9b5e5b5/Stim–Charitable-Food-Donations_Feb2012.pdf.aspx?ext=.pdf

FBLI (2015) 8th Institute, Food Bank Leadership Institute, Houston, Texas https://www.foodbanking.org/why-support-gfn/what-we-do/food-bank-leadership-institute/fbli-2015-inspiring-the-global-food-bank-community

FBLI (2016) *H-E-B/GFN Food Bank Leadership International Forum Institute* https://www.foodbanking.org/why-support-gfn/what-we-do/food-bank-leadership-institute

FDE (2016) Every meal matters. Press release. FoodDrinkEurope. http://www.fooddrinkeurope.eu/news/press-release/Every-Meal-Matters-European-food-bank-business-organisations-join-forces/

FDE (2017a) FoodDrinkEurope, Wikipedia https://en.wikipedia.org/wiki/FoodDrinkEurope

FDE (2017b) Guidelines: Every Meal Matters. FoodDrinkEurope http://www.fooddrinkeurope.eu/publications/every-meal-matters-food-donation-guidelines/

FEAD (2014) Poverty: Commission welcomes final adoption of new Fund for European Aid to the Most Deprived, Press Release, European Commission http://europa.eu/rapid/press-release_IP-14-230_en.htm

FEAD (2017) Fund for European Aid to the Most Deprived, Employment, Social Affairs and Inclusion, European Commission http://ec.europa.eu/social/main.jsp?catId=1089

FEBA (2016a) European Federation of Food Banks website https://www.eurofoodbank.eu/about-us/members

FEBA (2016b) History. European Federation of Food Banks https://www.eurofoodbank.eu/food-banking/history retrieved 30. 1. 17

FEBA (2016c) Review of the Fund for European Aid to the Most Deprived (FEAD), European Federation of Food Banks, Brussels 23 November

FEBA (2017a) Members. European Federation of Food Banks https://www.eurofoodbank.eu/about-us/members

FEBA (2017b) Governance. European Federation of Food Banks https://www.eurofoodbank.eu/about-us/governance

FEBA (2017c) FEBA's Team, European Federation of Food Banks https://www.eurofoodbank.eu/about-us/feba-staff

FEBA (2017d) Corporate or foundation partners, https://www.eurofoodbank.eu/get-involved/companies-foundations

FEBA (2017e) Food Waste http://www.eurofoodbank.org/poverty-waste/food-waste

FEBA (2017f) Values, Home Page. European Federation of Food Banks http://www.eurofoodbank.org/about-us/vision-mission-activities-values

FESBAL (2013) *La Labor de los bancos de alimentos en el año 2012.* [Online] Available: http://www.fesbal.org/

FF (2017) The Case for measuring UK Household Food Insecurity. The Food Foundation. http://foodfoundation.org.uk/event/the-case-for-measuring-uk-household-food-insecurity/

FIAN (2017a) FIAN International: Who we are. http://www.fian.org/who-we-are/who-we-are/

FIAN (2017b) FIAN International: monitoring and accountability http://www.fian.org/what-we-do/issues/monitoringaccountability/

FIAN-D (2013) Joint NGO Submission – UPR on the Federal Republic of Germany, FORUM MENSCHENRECHTE Berlin 2 October; Advocacy Note: FIAN International: Universal Periodic Review of Germany 3 May; see also U. Hausmann (2013)

Fidh (2013) Economic, Social and Cultural Rights Justiciable at the International Level. Press release, Worldwide Movement for Human Rights 5 May https://www.fidh.org.en

Fisher, A. (2017) *Big Hunger: The Unholy Alliance between Corporate America and Anti-Hunger Groups*. MIT Press

FLPC (2016) Legal Fact Sheet for Massachusetts Food Donation: Tax Incentives for Business – January 2016, Food Law and Policy Clinic, Harvard University http://recyclingworksma.com/wp-content/uploads/2015/07/Legal_Fact_Sheet_-_MA_Donation_Tax_Incentives-FINAL_RWF.pdf

FLW (2016) RELEASE: First-Ever Global Standard to Measure Food Loss and Waste Introduced by International Partnership http://www.wri.org/news/2016/06/release-first-ever-global-standard-measure-food-loss-and-waste-introduced-international www.FLWProtocol.org

Forbes (2017) Feeding America on Forbes Lists. The Hundred Largest US Charities, https://www.forbes.com.charities

Forcado, O., Sert, S. and Soldevila, V. (2015) New challenges against hunger and poverty: a food bank case study. Ch. 36 in Escajedo San-Epifanio, L. and De Renoblales Scheifler, M. *Envisioning a Future without Food Waste and Food Poverty*. Wageningen Academic Publishers

Forrester, V. (1999) *The Economic Horror*. Polity Press

Franciscus (2014) Message of His Holiness Francis for the Celebration of the World Day of Peace, Vatican, 1 January http://w2.vatican.va/content/francesco/en/messages/peace/documents/papa-francesco_20131208_messaggio-xlvii-giornata-mondiale-pace-2014.html

FRC (2016) Time to count the hungry: the case for a standard measure of household food insecurity in the UK, Food Research Collaboration – Economic & Social Research Council, The Food Foundation, Oxfam, Sustain and Oxford University http://foodfoundation.org.uk/wp-content/uploads/2016/07/Food-Poverty-workshop-report-1-05-04-16.pdf

Frost, R. (2016) The Road not Takenhttp://www.davidpbrown.co.ik/poetry/robert-frost-html

FSC (2017) Five big ideas for a better food system. Food Secure Canada, Montreal. https://foodsecurecanada.org/five-big-ideas

FT (2017) *Financial Times*, FT.com,/Lexicon http://lexicon.ft.com/Term?term=corporate-social-responsibility–(CSR)

F90 (2017) *Freedom 90 Charter*, Union of Food Bank and Emergency Meal Program Volunteers, Ontario, Canada http://www.freedom90.ca/charter.html

FUSIONS, (2016) *Estimates of European food waste Levels. Reducing food waste through social innovation*. Report to European Commission, FUSIONS, Stockholm http://ec.europa.eu/food/safety/food_waste_en

FW (2017) Fast Facts on Food Waste, FOODWISE http://www.foodwise.com.au/foodwaste/food-waste-fast-facts/

FWRA (2017) Food Waste Reduction Alliance http://www.foodwastealliance.org http://www.foodwastealliance.org/about-our-work/assessment/

Gambino, L. (2017) On the road with Sanders and Warren: Will the Democrats follow them to the left. *The Guardian* 27 August https://www.theguardian.com/us-news/2017/aug/27/bernie-sanders-elizabeth-warren-democratic-party-move-left

Gandy, J. and Greschner, S. (1989) Food Distribution Organisations in Metropolitan Toronto: a Secondary Welfare System? Working Papers in Social Welfare in Canada, Faculty of Social Work, University of Toronto.

Garthwaite, K. (2016) *Hunger Pains. Life inside Foodbank Britain*. Policy Press

Gentilini, U. (2013) Banking on Food: The State of Food Banks in High Income Countries, IDS working paper 415, Institute of Development Studies http://www.fdss.be/uploads/RevuepresseAA/2013/Wp415.pdf

Germann, J. (2009) The human right to food: "Voluntary guidelines" negotiations. In Atasoy, Y. *Hegemonic transition, the state and crisis in neoliberal capitalism.* Routledge

GFN (2006) FOOD, *Global Foodbanking Network Newsletter*, Winter https://www.foodbanking.org/wpcontent/uploads/2014/05/GFN_Winter06_English-1.pdf

GFN (2007) *Annual Report*, Global Food Banking Network http://www.foodbanking.org/wpcontent/uploads/2017/04/GFN2007annualreport.pdf

GFN (2011) One Billion Hungry and more than a Billion Tons of Food Wasted, Global Foodbanking Network https://www.foodbanking.org/one-billion-hungry-billiontons-food-wasted

GFN (2013a) Fact Sheet 4. 25. 2013, The Global FoodBanking Network https://www.oecd.org/site/agrfcn/GFN%20Fact%20Sheet%20a%20global%20non-profit.pdf

GFN (2013b) What is Food banking? The Global Foodbanking Network www.foodbanking.org

GFN (2015) GFN Marks a Decade of Global Food Banking. Newsletter

GFN (2016) Two Projects that transformed Food banking Globally. Newsletter. Global Foodbanking Network 18 February https://www.foodbanking.org/category/newsletters/

GFN (2016a) *Annual Report* The Global Foodbanking Network https://www.foodbanking.org/2016annualreport/

GFN (2017a) Food Banks Around the World, Global Foodbanking Network https://www.foodbanking.org/food-bank-resources/global-food-bank-community

GFN (2017b) Our People. Board of Directors. Global Foodbanking Network https://foodbanking.org/who-we-are/our-people/

GFN (2017c) Global Foodbanking Network Corporate and Foundation Partners https://www.foodbanking.org/who-we-are/our-corporate-and-foundation-partners/

GFN 24 August https://www.foodbanking.org/gfn-marks-a-decade-of-global-food-banking-as-gfn-approaches-its-10-year-anniversary-we-take-a-look-at-how-gfn-came-to-be/

GFSI (2016) Global Food Security Index, The Economist Intelligence Unit http://www.dupont.com/forms/dupont-food-security.html

G&M (2013) Ottawa rapped for hurting war on poverty, Ian Hunter, *Globe and Mail*, Toronto, 4 March.

Godoy, E. (2012) The waste mountain engulfing Mexico City, *The Guardian*, 9 January https://www.theguardian.com/environment/2012/jan/09/waste-mountain-mexico-city

Goldenberg, S. (2016a) America's food waste shame. As much as half of all food produce is thrown away. *The Guardian Weekly*, 15–21 July https://www.theguardian.com/environment/2016/jul/13/us-food-waste-ugly-fruit-vegetables-perfect

Goldenberg, S. (2016b) From field to fork: the six stages of wasting food. *The Guardian*, 14 July, 2016 https://www.theguardian.com/environment/2016/jul/14/from-field-to-fork-the-six-stages-of-wasting-food

Gonzáles Vaqué, L. (2015) Food loss and waste: some short- and medium-term proposals for the European Union. Ch. 1 in Escajedo San-Epifanio, L. and De Renobles Scheifler, M. (2015) *Envisioning a future without food waste and food poverty. Societal challenges*, Wageningen Academic Publishers

Graslie, S. (2013) Social Supermarkets A 'Win-Win-Win' for Europe's poor. *The Salt: NPR* 12 December http://www.npr.org/sections/thesalt/2013/12/11/250185245/social-supermarkets-a-win-win-win-for-europes-poor

Grover, E. (2017) Giving surplus supermarket food to charities will not solve hunger or waste problems, new paper claims *PHYS ORG* 26 January https://phys.org/news/2017-01-surplus-supermarket-food-charities-hunger.html

Guest, D. (1997) *The Emergence of Social Security in Canada*, 3rd edition. UBC Press

Gustavsson, J., Cederberg, C., Sonesson, U. and Emanuelsson, A. (2013) *The methodology of the FAO study: "global food Losses and Food Waste – extent, causes and prevention" – FAO, 2011.* SIK report No 857 The Swedish Institute for Food and Biotechnology.

Ha-joon Chang (2016) Austerity is based on lies. Owen Jones talks to Ha-joon Chang *The Guardian* 11 August https://www.theguardian.com/business/video/2016/aug/11/owen-jones-talks-toha-joon-chang-austerity-is-based-on-lies-video-interview

Hall, K., Guo, J., Dore, M. and Chow, C. (2009) The progressive increase of food waste in America and its environmental impact, *PLoS One*, 4(1), 1–6

Hamelin, A.-M., Beaudry, M., Habicht, J.-P. (2002) Characterization of household food insecurity in Québec: food and feelings. *Social Science and Medicine* 54(1), 119–132

Hänninen, S. and Karjalainen, J. (1994) Jälkisanat, in Hänninen, S. and Karjalainen, J. (eds) *Kirjeitä nälästä*. Stakes

Hansard (1988) Reply to the Speech from the Throne, Roy Romanow. Legislative Assembly of Saskatchewan 22 March http://docs.legassembly.sk.ca/legdocs/Legislative%20Assembly/Hansard/21L2S/88-03-22.pdf

Hausmann, U. (2013) Privatising the right to food: poverty, social exclusion and food banks in Germany. Ch. 10c *Right to Food and Nutrition Watch: Alternatives and Resistance to Policies that Generate Hunger.* FIAN International 2013 http://www.fian.org/fileadmin/media/publications_2015/Watch_2013_eng_WEB_final.pdf

H-E-B/GFN-FBLI (2015) Food Bank Leadership Institute, 2015 www.fao.org/save-food/news-and-multimedia/events/detail-events/en/c/278694/

H-E-B (2017) Here Everythings Better https://en.wikipedia.org/wiki/H-E-B

HLPE (2014) *Food losses and waste in the context of sustainable food systems.* A report by the High Level Panel of Experts on Food Security and Nutrition of the Committee on World Food Security, Rome 2014 www.fao.org/cfs/cfs-hlpe

Hodal, K. (2017) Bill Gates: Don't expect charities to pick up the bill for Trump sweeping aid cuts. *The Guardian* https://www.theguardian.com/global-development/2017/sep/13/bill-gates-foundation-dont-expect-pick-up-the-bill-for-sweeping-aid-cuts-trump

HRC (2016) UN Human Rights Council, *Report of the Special Rapporteur on the right to food on her mission to Poland, 27 December 2016, A/HRC/34/48/Add.1,* available at: http://www.refworld.org/docid/58ad970c4.html

HRC-B (2016a) *Report of the Working Group on the Universal Periodic Review –Belgium Addendum A/HRC/WG.6/24/BEL/3* 18–29 June

HRC-B (2016b) *Summary of stakeholders recommendations, Universal Periodic Review – Belgium,* Human Rights Council, Geneva 18–29 January http://www.ohchr.org/EN/HRBodies/UPR/Pages/BEIndex.aspx

HRC-USA (2015a) *National Report submitted in accordance with paragraph 5 of the annex to Human Rights Council resolution 16/21* – United States of America A/HRC/WG.6/22/USA/1* 4–5 May

HRC-USA (2015b) *Report of the Working group on the Universal Periodic review- United States of America A/HRC/WG.6/9/USA/1* 20 July

HRCWG/UPR (2013) *Advocacy Note, FIAN international suggested questions and recommendations: Universal Periodic Review of Germany.* Human Rights Council Working Group on the Universal Periodic Review 22 April – 3 May

Hunter, I. (2012) He's right: we should be ashamed. *Times Colonist*, Victoria BC 21 May http://www.accessibilitynewsinternational.com/hes-right-we-should-be-ashamed/

Hurtig, M. (1999) *Pay the Rent or Feed the Kids. The Tragedy and Disgrace of Poverty in Canada.* McClelland & Stewart Inc

ICESCR (1966/1976) *International Covenant on Economic, Social and Cultural Rights* http://www.ohchr.org/EN/ProfessionalInterest/Pages/CESCR.aspx

JAPAN (2011) *Japan – 3rd Periodic Report: Implementation of the International Covenant on Economic, Social and Cultural Rights due 2009, UN Economic and Social Council, E/C.12/JPN/3* 11 May http://tbinternet.ohchr.org/_layouts/treatybodyexternal/Download.aspx?symbolno=E%2fC.12%2fJPN%2f3&Lang=en

JF (2016) *Response to the List of Issues from the UN Committee on Economic. Social and Cultural Rights. Guidance and Background Information.* Just Fair www.just-fair.co.uk

Jochnick, C. (1999) Confronting the Impunity of Non-State Actors: New Fields for the Promotion of Human Rights, *Human Rights Quarterly* 21(1): 56–79 February

John Paul II (1991) Centesimus Annus, Encyclical Letter. 1 May http://w2.vatican.va/content/john-paul-ii/en/encyclicals/documents/hf_jp-ii_enc_01051991_centesimus-annus.html

Jones, O. (2014) 'Memo to Miliband: Britain's social order is bankrupt', Owen Jones, *Guardian Weekly*, 26 September https://www.theguardian.com/commentisfree/2014/sep/21/miliband-memo-britain-social-order-bankrupt

Jones, O. (2017) Most people in poverty are in work, earning their poverty. @OwenJones84

Jones, T. (2005) How much goes where. The corner on food loss. *BioCycle*, 2–3 Jul

JvH (2017) John van Hengel Wikipedia, the free encyclopedia https://en.wikipedia.org/wiki/John_van_Hengel retrieved 28. 1. 17

JvH (2006) Brief History of John van Hengel's Food Banking Concept (adapted from speech by Senator John McCain) https://www.feedingknowledge.net/

Kane, A. (2016) Australia's 7.5m tonnes of food waste; can 'ugly food' campaigns solve the problem? *The Guardian*, 6 June, 2016 https://www.theguardian.com/sustainable-business/2016/jun/06/australia-75m-tonnes-food-waste-ugly-food-solve-problem

Keenan, E. (2015) Shutting food bank first step in program to add respect to feeding hungry. *Toronto Star*, 8 February https://www.thestar.com/news/gta/2015/02/08/shutting-food-bank-first-step-in-program-to-add-respect-to-feeding-hungry-keenan.html

Keevers, L. (2010) *The Foodbank Story.* Received from Sarah Pennell of FoodBank Australia 2 February.

Kessl, F. (2016) "Charity Economy – a symbol of a fundamental shift in Europe" Unpublished paper, University of Duisburg-Essen

KHC (2017) Kick Hunger Challenge http://support.tasteofthenfl.com/site/PageServer?pagename=scorecard

Kleinhubbbert, G. (2014) Charities Struggle with Growing Demand 3 January.*Spiegel Online* http://www.spiegel.de/international/germany/german-food-banks-and-soup-kitchens-struggle-with-demand-a-941661.html

Klum, A. (2016) The roots of the first-ever right to food organization. FIAN International 10 October http://www.fian.org/en/news/article/the_roots_of_the_first_ever_right_to_food_organization/

Kontorravdis, E. (2016) *Report to the UN Committee on Economic, Social and Cultural Rights – On the Right to Food.* Nourish Scotland. 58th Session, UK 6th Periodic Review – Full Session, May

Kotz, N. (1984) The Politics of Hunger, *New Republic*, April 30, 19–23

Lambie-Mumford, H. (2013) 'Every town should have one': emergency food banking in the UK. *Journal of Social Policy*, 42(1), 73–89

Lambie-Mumford, H. (2017) *Hungry Britain. The Rise of Food Charity.* Policy Press

Lang, T. (2012) 'Food Matters: An integrative approach to food policy'. Paper based on talk to 'Mobilising the Food Chain for Health', Third meeting of the OECD Food Chain Analysis Network, 25–26 October, OECD Conference Centre, Paris.

Lang, T. (2013) Food waste is the symptom, not the problem. *The Conversation*, June 25 http://theconversation.com/food-waste-is-the-symptom-not-the-problem-15432

Lang, T. (2015) How to end Britain's destructive addiction to food banks. *The Conversation*, November 3 http://theconversation.com/how-to-end-britains-destructive-addiction-to-food-banks-50096

Lawrence, F. (2016) Agrochemicals and hyperintensive farming will never feed the world, *The Guardian*, 3 October https://www.theguardian.com/commentisfree/2016/oct/02/agrichemicals-intensive-farming-food-production-biodiversity

Leader-Post (2014) Eldon Anderson. Obituary. *Regina Leader-Post* http://www.legacy.com/obituaries/leaderpost/obituary.aspx?pid=171229383#sthash.MfAKpdZ5.dpuf

Leather, S. (1996) The Making of Modern Malnutrition. An overview of food poverty in the UK. The Caroline Walker Lecture

Lightman, E. (2003) *Social Policy in Canada*. Oxford University Press

Lightman, E. and Lightman, N. (2017) *Social Policy in Canada*. 2nd edition, Oxford University Press

Lindberg, R., Rose, N. and Caraher, M. (2016) The Human Right to Food in Australia. MUNPlanet blog https://www.munplanet.com/articles/food-security/the-human-right-to-food-in-australia

Lipinski, B. *et al.* (2013) "Reducing Food Loss and Waste". Working Paper, Installment 2 of *Creating a Sustainable Food Future*. Washington, DC: World Resources Institute. Available online at http://www.worldresourcesreport.org.

Loopstra, R. and Tarasuk, V. (2012) The relationship between food banks and household food insecurity among low-income Toronto families. *Canadian Public Policy* 38(4), 497–514

Loopstra, R. and Tarasuk, V. (2015) Food bank Usage is a Poor Indicator of Food Insecurity: Insights from Canada. *Social Policy & Society* 14(3), 443–455. doi:10.1017/S147476415000184

Lorenz, S. (2010) The German 'Tafel' – a sustainable way of dealing with food affluence? German Research Foundation http://www.soziologie.uni-jena.de/soziologie_multimedia/Downloads/LSRosa/Lorenz/The+German+Tafel.pdf

MacAskill, W. (2015) *Doing Good Better*. Avery

Macdiarmid, J., Lang, T. and Hanines, J. (2016) Down with food waste, *British Medical Journal*, 352(1), 138022 March

Mackay, R. (1995) Foodbank demand and supplementary assistance programmes: a research and policy case study. *Social Policy Journal of New Zealand* 2(5), 129–141

Mackenzie, H. (2017) Throwing Money at the Problem. 10 years of Executive Compensation. Ottawa: Canadian Centre for Policy Alternatives, 3 January https://www.policyalternatives.ca/newsroom/news-releases/ceo-pay-sets-new-record-study

Mackinnon, J. (2003) *Minding the Public Purse*. McGill-Queens University Press

Mann, A. (2016) Quoted in Peterson-Ward, J. in 'The right to food – and how 1.2 million Australians miss out', Division of Humanities and Social Sciences, University of Sydney 11 April

Martinko, K. (2017) What's the environmental impact of a loaf of bread? *TreeHugger* http://www.treehugger.com/green-food/whats-environmental-impact-loaf-bread.html

Maxwell, S. and Frankenberger, T. eds (1992) *Household Food Security: Concepts, Indicators, Measurements*. UNICEF and IFAD

McBain, S. (2015) Why are so many people using food banks? *New Statesman*, 30 March

McCain, J. (2003a) Brief History of John Van Hengel's Food Banking Concept https://www.feedingknowledge.net/02-search?

McCain, J. (2003b) Floor Statement of Senator John McCain commending the humanitarian work of John van Hengel http://.mccain.senate.gove/public/index.cfm/floor-statem ents?ID=F88ABD3B-C754-409C-B8BF-EOA99E593DD7

McGoey, L. (2015) *No Such Thing as a Free Gift: The Gates Foundation and the Price of Philanthropy*. Verso

McMahon, M. (2011) Global philanthropy versus food sovereignty. Department of Sociology, University of Victoria. Email communication 28 October

McPherson, K. (2006) Food Insecurity and the food bank industry: a geographical analysis of food bank use in Christchurch. Master's Thesis, University of Canterbury, Canterbury

MDGs (2015) *The Millennium Development Goals Report 2015*, United Nations http://www.un. org/millenniumgoals/2015_MDG_Report/pdf/MDG%202015%20rev%20(July%201).pdf

Melchior, M., Chastang, J.F., Falissard, B. et al. (2012) Food insecurity and children's mental health: a prospective birth cohort study. *PLoS ONE*, 7(12): e52615. doi:10.1371/journal. pone.0052615

MEX (2016) *Mexico – Combined Fifth and Sixth Periodic Reports due in 2012* , UN Economic and Social Council, E/C.12/MEX/5–6, 8 June http://tbinternet.ohchr.org/_layouts/treaty bodyexternal/Download.aspx?symbolno=E%2fC.12%2fMEX%2f5-6&Lang=en

Monbiot, G. (2016) Neoliberalism – the ideology at the root of all our problems. *The Guardian* 15 April http://www.theguardian.com/books/2016/apr/15/neoliberalism-ideo logy-problem-george-monbiot

Moore, S. (2016), Politicians have got what they wanted: more UK workers in poverty than ever before. *The Guardian*, 7. 12. 16 https://www.theguardian.com/commentisfree/ 2016/dec/07/politicians-have-got-what-they-wanted-more-uk-workers-in-poverty-than-ever-before

Mourad, M. (2015) From food waste to wealth: valuing excess food in France and the USA. Ch. 7 in Escajedo San-Epifano, L. and De Renobales Scheiffler, M. *Envisioning A Future Without Food Waste and Food Poverty*. Wageningen Academic Publishers

Moyn, S. (2014) The Future of Human Rights. *SUR – International Journal On Human Rights*, 11(2), Jun./Dec. SSRN: https://ssrn.com/abstract=2550376 https://papers.ssrn. com/sol3/papers.cfm?abstract_id=2550376

NOR (2012) *Norway, Fifth Periodic Report: Implementation of the International Covenant on Economic, Social and Cultural Rights due 2010*, UN Economic and Social Council, E/C.12/ NOR/5 29 October http://tbinternet.ohchr.org/_layouts/treatybodyexternal/Download. aspx?symbolno=E%2fC.12%2fNOR%2f5&Lang=en

Norden (2015a) *New Nordic Study: Food banks have a big unused potential to minimize food waste*. Nordic Co-operation. Report for Nordic Council of Ministers 06. 02. 2015 http://www. norden.org/en/news-and-events/news/new-nordic-study-food-banks-have-a-big-unused-potetianl-to-minimize-food-waste

Norden (2015b) *Food Redistribution in the Nordic Region. Experiences and results from a pilot study*. Report for the Nordic Council of Ministers http://norden.diva-portal.org/smash/ get/diva2:784307/FULLTEXT01.pdf

NP (2011) Vancouver is the world's most livable city for a fifth straight year: survey. *National Post* http://news.nationalpost.com/news/canada/vancouver-is-the-worlds-most-livable-ci ty-for-a-fifth-straight-year-survey

NS (2014) Our Common Wealth of Food. Report – Nourish Conference 2014. Glasgow. Nourish Scotland www.nourishscotland.org/wp-content/uploads/2015/02/Conference-Report-2014.pdf

NYT (2017) Ben Carson calls Poverty a 'State of Mind', Igniting a Backlash. *New York Times*, 25 May https://www.nytimes.com/2017/05/25/us/politics/ben-carson-poverty-hud-state-of-mind.html?_r=0

O'Brien, M. (2014) Privatizing the Right to Food: Aotearoa/New Zealand. Ch.8 in Riches, G. and Silvasti, T. *First World Hunger Revisited*. Palgrave Macmillan

OC (2013) Oblate Communications 532, The Missionary Oblates of Mary Immaculate, April http://omiworld.org/en/information/532/april-2013/?Page=6

OECD (2017a) About the OECD http://www.oecd.org/about/

OECD (2017b) *Report of the Chair of the Working Group on the Future Size and Membership of the Organisation to Council*. Meeting of the OECD Council at Ministerial Level. Paris 7–8 June http://www.oecd.org/mcm/documents/C-MIN-2017-13-EN.pdf

OHCHR (2013a) *Mission to Canada. Report of the Special Rapporteur on the Right to Food, Olivier de Schutter, Human Rights Council, General Assembly 24 December A/HRC/22/50/ Add.1*

OHCHR (2013b) *Working visit to Mexico by the Special Rapporteur on the right to food*. UN Office of the High Commission for Human Rights, Geneva 14–15 November

OHCHR (2017a) *Special Rapporteur on the Right to Food. Office of the High Commissioner for Human Rights* 6 June http://www.ohchr.org/EN/Issues/Food/Pages/FoodIndex.aspx

OHCHR (2017b) What we do. Office of the Hugh Commission for Human Rights, United Nations http://www.ohchr.org/EN/AboutUs/Pages/WhatWeDo.aspx

OHCHR (2017c) Universal Periodic Review. Human Rights Council http://www.ohchr. org/EN/HRBodies/UPR/Pages/UPRMain.aspx

OHCHR (2017d) Documentation by countries. Human Rights Council http://www. ohchr.org/EN/HRBodies/UPR/Pages/Documentation.aspx

OHCHR (2017e) Periodic Reports. Committee on Economic Social and Cultural Rights, Office of the High Commission for Human Rights, Geneva 2 June http://tbinternet. ohchr.org/_layouts/treatybodyexternal/TBSearch.aspx?Lang=en&TreatyID=9&DocTyp eID=29

OHCHR-SRRF (2017) Special Rapporteur on the Right to Food. UN Office of the High Commission on Human Rights, Geneva. http://www.ohchr.org/EN/Issues/Food/Pages/ FoodIndex.aspx

O'Hearn, T. (2004) Foodbank Victoria: In the Right Place between Demand and Supply. *Parity*, 17(3), 8–9

O'Hogan, C. (2017) Do ut des, translation, Dr. Cillian O'Hogan, Department of Classical, Near Eastern, and Religious Studies, UBC, personal communication. 8 May

O'Neill, O. (1986) *Faces of Hunger. An Essay on Poverty, Justice and Development*. Allen & Unwin

OSNPPH (2015) Position Statement on Responses to Food Insecurity, Ontario Society of Nutrition Professionals in Public Health, 26 November www.osnpph.on.ca

Owen, J. (2014) Busy Nottingham food bank to close in protest at harsh council cuts. *Independent News*. http://www.independent.co.uk/news/uk/politics/busy-nottingham- food-bank-to-close-in-protest-at-harsh-council-cuts-9885641.html

Oxfam (2013) *Behind the Brands. Food Justice and the 'Big 10' food and beverage companies* 166 Oxfam Briefing Paper 28 February https://www.oxfam.ca/sites/default/files/imce/ btb-behind-the-brands-report.pdf

Oxfam (2017) *An Economy for the 99%*. Oxfam Briefing Paper. January https://www.oxfam. org/en/research/economy-99

Pérez de Armiño, K. (2014) Erosion of rights, uncritical solidarity and food banks in Spain, Ch. 10 in Riches, G. and Silvasti, T. *First World Hunger Revisited*. Palgrave Macmillan

PFC (1980) *The Land of Milk and Money*, National Report of the People's Food Commission. Canada: Between the Lines

PFPC (2011) *Resetting the Table. A People's Food Policy for Canada*. Food Secure Canada https:// foodsecurecanada.org/sites/foodsecurecanada.org/files/FSC-resetting2012-8half11-lowres- EN.pdf

Placido, S. and Rietberg, P. (n.d.) *Food Banks in Amsterdam: More than Just Your Daily Bread*. Humanity in Action http://www.humanityinaction.org/knowledgebase/320-food-banks-in-amsterdam-more-than-just-your-daily-bread

Poppendieck, J. (1997) The USA: hunger in the land of plenty. Ch. 6 in Riches, G. *First World Hunger. Food Security and Welfare Politics*. Macmillan

Poppendieck, J. (1998) *Sweet Charity. Emergency Food and the End of Entitlement*. Viking Press

Poppendieck, J. (2014a) Food assistance, hunger and the end of welfare in the USA. Ch. 13 in Riches, G. and Silvasti, T. *First World Hunger Revisited: Food Charity or the Right to Food*. Palgrave Macmillan

Poppendieck, J. (2014b) *Breadlines Knee-Deep in Wheat. Food Assistance in the Great Depression*. 2nd edition, University of California Press

Power, E. (2011) Canadian Food Banks. Obscuring the reality of hunger and poverty. *Food Ethics*, 6(4), 18–20 www.foodethicscouncil.org

Powers, J. (2015) Food is a right, not an act of charity. *Why Hunger: Hunger Action Network*, New York 13 October https://whyhunger.org/connect/item/3025-the-right-to-food

Powers, J. (2015b) Surplus and Suffering: Book Review of "First World Hunger Revisited". *WhyHunger* 16 February https://whyhunger.org/surplus-suffering-first-world-hunger-revisted/

Powers, J. and Cohen, A. (2016) America's Food Banks say Charity won't end Hunger, Special Report. *WhyHunger* New York 21 January @WhyHunger, @foodandfury

Pretty, J. (2004) We are what we eat. *New Scientist*, 9 October www.newscientist.com

PROOF (2014) *Household Food Security in Canada*. PROOF Report Highlights 2014 Numbers https://foodsecurecanada.org/resources-news/news-media/proof-report-2014

ReFED (2016) "A Roadmap to Reduce U.S. Food Waste by 20 Percent" https://www.refed.com/downloads/ReFED_Report_2016.pdf

René, P.-M. (2106) 19 million tons of food are wasted in Mexico every year. *EL Universal* http://www.eluniversal.com.mx/articulo/english/2016/02/11/

RFU-UNFAO (2008) Right to Food Unit, Food and Agricultural Organization, Rome

Riches, G. (1986) *Food Banks and the Welfare Crisis*. Canadian Council on Social Development

Riches, G. ed. (1997) *First World Hunger: Food Security and Welfare Politics*. Macmillan

Riches, G., Buckingham, D., Macrae, R. and Ostry, A. (2004) *Right to Food Case Study: Canada*. Rome United Nation's Food and Agricultural Organisation IGWG RTFG/INF 4/APP2

Riches, G. and Graves, J. (2007) Let them eat starch, *The Tyee* 28 August https://thetyee.ca/Life/2007/08/28/FoodLines/

Riches, G. and Silvasti, T. eds, (2014) *First World Hunger Revisited. Food Charity or the Right to Food?* Palgrave Macmillan

Riches, G. and Tarasuk, V. (2014) Canada: thirty years of food charity and public policy neglect. Ch. 4 in Riches, G. and Silvasti, T. *First World Hunger Revisited. Food Charity or the Right to Food?* Palgrave Macmillan

Rieff, D. 2015 *The Reproach of Hunger. Food, Justice and Money in the Twenty-first Century*. Simon & Schuster

Robertson, A., Brunner, E. and Sheiham, A. (1999) Food is a political issue. Ch 9 in Marmot, M. and Wilkinson, R.G. eds *Social Determinants of Health*. OUP

Robinson, M. (2004) 'Ethics, Globalization and Hunger: In Search of Appropriate Policies', Keynote Address, Cornell University, 18 November

Roff, R. (2016) *The Farmers' Food Donation Tax Credit misses the mark in food security*. Policy Note, Canadian Centre for Policy Alternatives – BC, Vancouver http://www.policynote.ca/the-farmers-food-donation-tax-credit-misses-the-mark-in-food-security/

Ronson, D. and Caraher, M. (2016) Food banks: big society or shunting yards? Successful failures. Ch. 8 in Caraher, M. and Coveney, J. (eds) *Food Poverty and Insecurity: International food Inequalities*. Springer

Rotondaro, V. (2015) Report details massive financial cost of American hunger and food insecurity. *National Catholic Report*, November 15 https://www.ncronline.org/blogs/ ncr-today/report-details-massive-financial-costamerican-hunger-and-food-insecurity

RTFC (2017) Right to Food Coalition (Australia) *The Human Right to Food* https://right tofoodcoalition.files.wordpress.com/2016/04/human-right-to-food-position-statement-170416.pdf https://righttofood.com.au

Salonen, A. (2016) 'Food for the Soul or the Soul for the Food'. Users perspectives on religiously affiliated food charity in a Finnish city. Academic Dissertation, Faculty of Theology, University of Helsinki, Finland 12th October

Schlosser, E. (2001, 2012) *Fast Food Nation*, Mariner Books Houghton, Mifflin Harcourt

Scott, C. (2016) Love Food Hate Waste campaign launched. *Newshub* 31 May http://www. newshub.co.nz/home/new-zealand/2016/05/love-food-hate-waste-campaign-launched. html

Scott-Thomas, C. (2013) Low food industry wages are a British disease, says food policy expert. *Food Navigator.com* http://www.foodnavigator.com/Market-Trends/Low-food-industry-wages-are-a-British-disease-says-food-policy-expert

SDGs (2015a) *Transforming Our World – the 2030 Agenda for Sustainable Development*. United Nations https://sustainabledevelopment.un.org/topics/sustainabledevelopmentgoals

SDGs (2015b) Goal 12, Social Development Goals, United Nations http://www.undp.org/ content/undp/en/home/sustainable-development-goals/goal-12-responsible-consumption-and-production.html

Selke, S. (2013) The rise of food banks in Germany is increasing the commodification of poverty without addressing its structural causes. EUROPP-European Politics and Policy: LSE http://bit.ly/12ITfWT

Selke, S. (2015) *Schamland, Die Armut mitten unter uns*. Ullstein Buchverlage GmbH

Sert, S., Garrone, P., Melacini, M. and Perego, A. (2015) Surplus food redistribution fo social purposes: analysis of critical success factors. Ch. 8 in Escajedo San-Epifanio, L. and De Renobales Scheifler, M. *Envisioning a Future without Food Waste and Food Poverty*. Wageningen Academic Publishers

SFC (2017) Core Principles. Scottish Food Coalition http://www.foodcoalition.scot/our-core-principles.html

Shimmin, C. and Tarasuk, V. (2015) The ugly truth: Many Canadians didn't have enough food this year. *The Globe and Mail* 23 December https://www.theglobeandmail.com/op inion/the-ugly-truth-many-canadians-didnt-have-enough-food-this-year/article27920948/

Silvasti, T. and Karjalainen, J. (2014) Hunger in a Nordic welfare state: Finland. Ch. 6 in Riches, G. and Silvasti, T. *First World Hunger Revisited*. Palgrave Macmillan

Silvasti, T. and Riches, G. (2014) Hunger and food charity in rich societies: what hope for the right to food? Ch. 14 in Riches, G. and Silvasti, T. *First World Hunger Revisited*. Palgrave Macmillan

Silvasti, T. and Tikka, V. (2015). *National Report: Finland*. Transmango. National media reports on Food and Nutrition Security. Retrieved from http://www.transmango.eu/ userfiles/project%20reports/d2.2%20p7%20finland

Smithers, R. (2013) Asda unveils food bank pledge for surplus stock. *The Guardian* 3 June https:// www.theguardian.com/money/2013/jun/03/asda-food-bank-pledge-surplus-stock

Solnit, R. (2016) *Hope in the Dark. Untold Histories. Wild Possibilities*. Canongate

Soma, T., Li, B. and Kranenburg, D. (2016) Food Systems Lab Design Brief, Pierre Elliott Trudeau Foundation, Toronto Nov 24 and 25

Soskis, B. (2014) The Importance of Criticizing Philanthropy. *The Atlantic* May 12 https://www.theatlantic.com/business/archive/2014/05/the-case-for-philanthropy-criticism/361951/

Steyaert, J. (2014) John van Hengel & food banks: philanthropy for the hungry. *History of Social Work*, University of Antwerp http://www.historyofsocialwork.org/eng/details.php?cps=18

Stuart, T. (2009) *Waste. Uncovering the Global Food Waste Scandal*. Penguin Books

Supiot, A. (2001) *Employment: Changes in Work and the Future of Labour Law in Europe*. OUP

Tang, F. (2017) The German "Tafel" feeding the needy. *Food Explorers* http://www.foodexplorers.net/tafel-feed-needy

Tarasuk, V. (2016) The Full Story of Food (In)security in Canada. PROOF http://proof.utoronto.ca/resources/fact-sheets/#monitoring

Tarasuk, V. and Beaton, G. (1999) Women's dietary intakes in the context of household food insecurity. *The Journal of Nutrition*, 129(3), 672–679

Tarasuk, V. and Eakin, J. (2003) Charitable food assistance as symbolic gesture: an ethnographic study of food banks in Ontario. *Social Science and Medicine*, 56(7), 1505–1515

Tarasuk, V. and Eakin, J. (2005) Food assistance through "surplus" food: Insights from an ethnographic study of food bank work. *Agriculture and Human Values* 22, 177–186. doi:10.1007/s10460-004-8277-x

Tarasuk, V., Cheng, J., de OliveiraC., Dachner, N., Gundersen, C. and Kurdyak, P. (2015) Health care costs associated with household food insecurity in Ontario. *Canadian Medical Association Journal* http://www.cmaj.ca/content/ early/2015/08/10/cmaj.150234

Tarasuk, V., Mitchell, A. and Dachner, N. (2014). *Household food insecurity in Canada, 2012*. Toronto: Research to identify policy options to reduce food insecurity (PROOF). Retrieved from http://nutritionalsciences.lamp.utoronto.ca/

Tarasuk, V., Mitchell, A. and Dachner, N. (2016). *Household food insecurity in Canada, 2014*. Toronto: Research to identify policy options to reduce food insecurity (PROOF). Retrieved from http://proof.utoronto.ca

Taylor, A. and Loopstra, R. (2016) Too Poor to Eat. Food insecurity in the UK. Food Insecurity Briefing, The Food Foundation, May www.foodfoundation.org.uk

TC (2012) "Even some food banks feel penalized by NHL lockout", *Times Colonist (Victoria)*, 14 December www.timescolonist.com/life/even-some-food-banks-feel-penalized-by-nhl-lockout-1.27738 20 CBC

TGG (2016) Second Harvest – Japan's First Nationwide Food Bank, Karianne. *The Good Globe*. https://thegoodglobe.com/second-harvest-Japan

The Press (2013) Councillor in attack on food bank. *The Press*, 3 January http://www.yorkpress.co.uk/news/10138097.Councillor_in_attack_on_food_bank/

Townsend, P. (1979) *Poverty in the United Kingdom. A Survey of Household Resources and Standards of Living*. University of California Press.

Track, L. (2015) *Hungry for Justice: Advancing a Right to Food for Children in BC*. British Columbia Civil Liberties Association, Vancouver

TT (2013) Briefing for debate on food banks. The Trussell Trust, 18 December https://www.trusselltrust.org/wp-content/uploads/sites/2/2015/06/Foodbank-Briefing-for-MPs.pdf

TT (2017) https://www.trusselltrust.org/

Tyee (2007) On-line commenter, Let them eat starch. *The Tyee*, British Columbia https://thetyee.ca/Life/2007/08/28/FoodLines/

UN(United Nations) General Assembly (2004) *Report of the Special Rapporteur on the right to food, Jean Ziegler. 27 September 2004, A/59/385*

UN Fact Sheet (2015) *We can end poverty. Millennium Development Goals and Beyond 2015*. http://www.un.org/millenniumgoals/pdf/Goal_1_fs_pdf

UNESC (1999) *General Comment 12, The Right to Food , United Nations Economic and Social Council E/C.12/1999/5 12 May 1999* http://www.fao.org/fileadmin/templates/right tofood/documents/RTF_publication EN/General_Comment_12_EN.pdf

USDA ERS (2016a) Definitions of Food Security. Economic Research Service, United States Department of Agriculture https://www.ers.usda.gov/topics/food-nutrition-assistance/food-security-in-the-us/definitions-of-food-security.aspx

USDA ERS (2016b) Measurement. Economic Research Service, United States Department of Agriculture https://www.ers.usda.gov/topics/food-nutrition-assistance/food-security-in-the-us/measurement.aspx

USDA ERS (2016c) Coleman-Jensen, A., Rabbitt, M.P., Gregory, C.A. and Singh, A. *Household Food Security in the United States in 2015*, ERR-215, U.S. Department of Agriculture, Economic Research Service, September 2016. https://www.ers.usda.gov/web docs/publications/79761/err-215.pdf?v=42636

Uttley, S. (1997) Hunger in New Zealand: a question of rights. Ch 4 in Riches, G. (ed.) *First World Hunger. Food Security and Welfare Politics*. Macmillan

Van Beuren, G. (2013) Should we have an enforceable right to food? *UK Human Rights Blog*, 18 October https://ukhumanrightsblog.com/2013/10/18/should-we-have-an-enforceable-right-to-food-professor-geraldine-van-bueren/

Van der Horst, H., Pascucci, S. and Bol, W. (2014) The "dark side" of food banks? Exploring emotional responses of food bank receivers in the Netherlands. *British Food Journal*, 116(9), 1506–1520, doi:10.1108/BFJ-02-2014-0081

VCMI (2014) '27 Billion Revisited.' The Cost of Canada's Annual Food Waste. Value Chain Management, Inc http://vcm-international.com

Visser, M. (2000) *Much Depends on Dinner*. Harper Perennial Canada

Vozoris, N. and Tarasuk, V. (2003) Household food insufficiency is associated with poorer health. *J Nutr*, 133(1), 120–126.

Wells, R. and Caraher, M. (2014) UK print media coverage of the food bank phenomenon: from food welfare to food charity, *British Food Journal* 116(9), 1426–1445

Williams, Z. (2014) As children starve, where's the state? *The Guardian*, 9 December https://www.theguardian.com/commentisfree/2014/dec/09/children-starve-state-food-banks

Wilson, J. (1997) Australia: Lucky country/hungry silence. Ch. 2 in Riches, G. (ed) *First World Hunger. Food Security and Welfare Politics*. Macmillan

Winne, M. (2008) *Closing the Food Gap. Resetting the Table in the Land of Plenty*. Beacon Press

Winson, A. (1993) *The Intimate Commodity*. Garamond Press http://www.emeraldinsight.com/doi/abs/10.1108/BFJ-03-2014-0123

WNC-NZFWA (2015) Waste Not Consulting. *New Zealand Food Waste Audits*, prepared for WasteMinz http://www.wasteminz.org.nz/pubs/new-zealand-food-waste-audits-2014-2015/

Wood, Z. (2016) Tesco food waste rose to equivalent of 119m meals last year. *The Guardian* https://www.theguardian.com/business/2016/jun/15/tesco-food-waste-past-year-equivalent-119-million-meals

WRAP (2017a) *Estimates of Food Surplus and Waste Arisings in the UK*. Waste & Resources Action Programme, WRAP UK http://www.wrap.org.uk/sites/files/wrap/Estimates_in_the_UK_Jan17.pdf

WRAP (2017b) The Courtauld Commitment http://www.wrap.org.uk/node/14507

WRI (2017) World Resources Institute http://www.wri.org//

Ziegler, J., Golay, C., Mahon, C. and Way, S.-A. (2011) *The Fight for the Right to Food: Lessons Learned*. Palgrave Macmillan

INDEX